THE WILD LIFE OF THE DOMESTIC CAT

Roger Tabor, M.I.Biol., M.Phil., F.L.S., is a professional biologist, freelance broadcaster and writer. He first found himself on television and radio in the early 1970s as a result of his motorway verge behavioural ecology research as biologist to the North East London Polytechnic's Motorway Research Project. His particular interests are the associations between man and wildlife, as in urban ecology, and in recent years he has been prominent in the areas of cat behaviour and ecology. His involvement with urban wildlife extends into his broadcasting with his weekly London wildlife slot on LBC; in contrast he has a fortnightly slot on BBC 1 in arable East Anglia. He has appeared on a wide range of programmes. Past editor of 'Country-side', he is currently scientific officer of the British Naturalists' Association. He leads wildlife parties to India, Nepal and South America, and has travelled widely watching cats both big and small.

THE WILD LIFE OF THE DOMESTIC CAT

Roger Tabor

ARROW BOOKS

Arrow Books Limited
17-21 Conway Street, London W1P 6JD

An imprint of the Hutchinson Publishing Group

London Melbourne Sydney Auckland
Johannesburg and agencies throughout
the world

First published by Arrow 1983
Reprinted 1984

Set in Linotron Sabon by
Rowland Phototypesetting Ltd
Bury St Edmunds, Suffolk

Made and printed in Great Britain
by The Guernsey Press Co Ltd
Guernsey C.I.

Colour Plates printed by
Balding & Mansell Ltd
Wisbech, Cambridgeshire

ISBN 0 09 931210 7

To my wife Rachel, and to our cat who eats butterflies, so despite her sex has to be called 'Mr Jeremy Fisher', her 'effanineffable, deep and inscrutable singular Name'. (Apologies to Beatrix Potter and T. S. Eliot.)

Credits:

Illustrations : Eugene Fleury and author
Photograph : Galapagos kitten and Galapagos Hawk with cat prey by John Budris.
All other photographs and the scanning electron micrographs by the author.

The following illustrations were redrawn after the authors below:

Page 31 after Ulmer, Haupt and Hicks (1971)
Page 47 after Leyhausen (1979)
Page 60 after MacDonald and Apps (1978)
Page 63 after Dards (1981)
Page 71 general species curves after McNab (1963), cat blocks by author
Page 79–80 after MacDonald and Apps (1978)
Page 151 after Thornton (1971)

Contents

Foreword

'Once Cattes were all wilde, but afterward they retyred to houses, wherefore there are plenty of them in all countries'

Edward Topsell (17th century)
The Historie of Foure-footed Beasts

Topsell's quaintly expressed comment today encapsulates much of the story of the domestic cat. Domestication has been a successful survival ploy for the cat. Cats have exploited a freely available food source by associating with man and consequently can have small home ranges and territories and so build up to huge numbers. They have also gained wide dispersion across the globe. None the less they have not been enclosed and restrained like other domestic animals, they have remained essentially free-living. Although cats enjoy living with man, they easily revert to the wild or feral existence. House cats, strays and feral cats, town and country, lead surprisingly similar lives despite the obvious difference of their relationship to man. Although all are domestic cats (*felis catus)*, each gives insight into the wildlife of the other.

Harrison Weir, Instigator of the first cat show at the Crystal Palace.

HISTORICAL INTRODUCTION
Domestication of Demon and Deity

Domestication of man . . . then cat!

Although recognisable man may have emerged into a grassland habitat in Africa his initial build-up of numbers forced him into the bleaker more northern conditions of the recently dawned ice age. An existence of a semi-nomadic hunter-gatherer of a small, but apparently relatively stable population, lasted with the resurgence of ice some hundreds of thousands of years.

Following the withdrawal of the ice, man gradually adapted again to his new climate and by the Neolithic period the wolf had, by association with hunting and then herding parties, become domesticated into the dog. However the huge change that allowed man to become settled, and therefore in a better position to develop artifacts that would be inconvenient to a mobile life, thus to increase his arts, technology and allow more animals to associate with him domestically, was the cultivation of grain. This was as a result of man's collection methods, and the natural hybridisation occurring between grasses, whereby the seeds were less able to fall from the ears of the new wheat.

As with most stories of domestication man's hand was only partly responsible and if we were looking from outside of ourselves as we do at other species, we would not think in terms of 'domestication' for we would see a relationship of benefit to both partners, a symbiotic dependence. The new grasses exploit man as their dispersal agent and gain a widespread distribution, while in man features that we would recognise as among the general trend of domestication followed, such as the shortening of the facial part of the skull.

Following man's domestication, his more settled existence naturally attracted less timerous species, such as crop-robbing cattle. The storage of grain led to the attraction of domestic wild (as opposed to domesticated species) species of seed eating birds and rodents. The development of villages and small towns produced quantities of accumulating refuse. From what we have found of today's feral or 'gone wild' cat, these conditions were now ideal to attract the emerging species of what we now recognise as the domestic cat.

The origins of the cat

The origin of the cat is still shrouded in mystery. This is perhaps apt for an animal that stepped from the wild into apparent domestication as a god and was carefully mummified on death.

Our present domestic cat as a member of the cat family is Felis, but it is sometimes referred to as F. catus, following the lead of Linnaeus, father of zoological classification, and sometimes it is called F. domesticus.

The existing evidence points to the emergence of the domestic cat from the Middle East, and Egypt in particular. Of the many small sized wild cats in the world, the most likely progenitors are those whose range crosses the area, Felis sylvestris and Felis libyca, the European and African wild cats, and possibly Felis chaus, the Jungle cat.

The European and African wild cats (F. sylvestris and F. libyca) were once considered two distinct species, but a gradual transition of coat colour and length has been demonstrated with the previously believed 'true' types being at extremes of the range.

The earliest mammalian carnivores (credonts) are thought to have emerged in the Palaeocine and Eocene periods. In evolutionary terms the families of Canidae and Felidae diverged during the Eocene (40–60 million years ago) with a move away from a generalised hunter. The dawn of the oligocene period revealed a civet-like animal, followed by the more cat-like animals of the Miocene age (10–25 million years ago). Although sabre-toothed cats were an early branch, forking during the Miocene age, most of the modern cats seem to have separated during the late Miocene to Pliocene ages (1–15 million years ago). Ones similar to today's big cats appeared in the early period, and this may have been due to the formation of open grassland habitat with suitable prey species. The middle and late Pleistocene were the periods of the Ice Ages and it was during these that the smaller wildcats were firmly in evidence.

Development of the cat family

As cats developed the relative skull length shortened, allowing the jaws an increased shearing force. Both hearing and vision is more developed in the modern cats than in dogs, but the cat's sense of smell is probably not quite as acute.

The wild cats range in size from the very small with body and head average lengths of 30–50cms up to the very large tiger and lion at around 200cms long. Felis silvestris (lybica) on this basis, although being small, are not grouped with the smallest cats.

The cat family is normally seen as having little social structure, with most representatives living solitary lives. It is often stated that the lion is the only truly social felid, and attention has recently been focussed on the lion's social structure of mother with daughters, which is very different from the pack hunting structure of the canids. However, it is now becoming clear that the domestic cat does have an inherent social structure, but not one that is linked to group hunting, for the image of 'the cat who hunts alone' is still intact.

Social hunting units are as much a reflection of the prey as of the predator. Canids employ it frequently as the only way of overcoming large prey. It is usually used against members of large herds and this is as true of lions as wolves. Communal feeding affords the significant advantage of preventing other large predators and scavengers from pinching the large kill! These are not problems encountered by the domestic cat in domestic or feral state, so little pressure has existed for communal hunting. Even the lion in habitats away from the open savannah is less under pressure by the above conditions and easily loses its group hunting identity.

The domestic cat, when it becomes self-sufficient in a feral existence, has a pattern of behaviour and social structure apparently distinct from one of its believed progenitors, the wild cat (Felis sylvestris). When the role of attached human beings is considered it seems that the domestic cat on 'going feral' does not really develop a new social structure if people are thought of as 'surrogate cats', for the similarities imply just a shift in group identity. We do not know which is the 'natural state' of the domestic cat i.e. whether the cat's wild ancestors were lured initially into captivity, or whether the new cat appeared in the wild and then adopted man. What is clear however is that the domestic cat has lived both in the true domestic and in the feral state successfully for thousands of years.

Clues to the origin of the domestic cat

Mummified cats

In 1889 at Beni Hassan in Central Egypt a very large number of 'mummified' cats were discovered. Each of these ancient cats had been separately embalmed and laid in rows in an underground cavern. As the find was of the order of some hundreds of thousands, they seemed instantly too available to hold a price and consequently most were sold to local farmers as fertilizer. One lot of nineteen and a half tons found its way to Liverpool by the steamer 'Pharos and Thebes'. Apparently the auction that followed had only been advertised to the agricultural market and the mummified cats were sold at between £3.50 and £4.25 per ton. The auctioneer used one of the cats' heads as a gavel! Only a very few were selected out for museum use.

As the definite origin of the domestic cat is still not fully known, for such a stockpile of evidence of the probable emergence of the domestic cat to have been carefully preserved and survived thousands of years, only to be ground to dust a matter of days before a record could have been made, must rank fairly high among the acts of wanton vandalism executed in the interest of short term small monetary gain.

It is fortunate that a few of the mummified cats remain and at a time of Victorian collections this is perhaps the most we could expect. But the almost total loss of such a number that could have provided mathematically significant data is, to say the least, unfortunate. However, fortunately in 1907 the British Museum was presented with 192 cats from Gizeh (Egypt) dating from around 600–200BC. The ancient Egyptian cats were thought for many years, following an investigation by Ehrenberg in 1833, to be in three forms. One was the larger Felis chaus (or Jungle cat) and the two other forms were both smaller cats. However in 1952 it was shown that these smaller mummified types in the British Museum collection were really only of one sort and from their skulls seemed similar to Felis libyca (the African wildcat). This was believed to be a domestic variant to be known as Felis libyca bubastis and from contemporary paintings and figures it was suggested that it was ginger-coloured with long ears and legs and topped by a long ringed tail.

As the number of F. chaus mummified was very small compared to F. libyca bubastis a case could apparently be made for this latter to be the ancestor of your cat, Felis catus. Nonetheless in the skull measurements taken it is clear that although similar, the mummified cats form a distinct group from today's African wildcats and domestic cats, and if anything modern domestic cats' measurements are more directly comparable to F. libyca than to the larger F. libyca bubastis.

As the sexes of the mummies are unknown, the difference to F. libyca would become less if for any reason the Egyptians had mummified only male cats or separated the burial chambers on a sex basis. The British Museum cats come from a period when the cat had already been co-habiting with man for a long period.

The date of domestication (which may have been gradual) has been suggested as being as early as 3000BC, but it is normally agreed as occurring some time over the following 1500 years. The Egyptians are believed to have jealously guarded their cats to prevent the cats reaching other kingdoms. If consideration is given to the current fluidity of the cat population it is possible that greater effective restrictions were then placed on the cats. It is even possible that restriction of mixed wild populations to temple premises allowed the appearance of a domestic hybrid. If such were the case the suggestion that the domestic cat emerged from a cross between the Euro/African wildcat, F. libyca (silvestris), and the Jungle cat, F. chaus, could have foundation, instead of being a variant of just F. libyca (silvestris). F. chaus were apparently kept by the Egyptians and the hybrid at least of chaus and catus is not only fertile but larger and with longer legs than catus, and further has the advantage of being of tame disposition. These make a cross a strong contender for the original domestic cat. Certainly at the time of this early melting pot it is possible that once bubastis had appeared by either a variant of libyca alone, or resultant of chaus/libyca cross, as well as continued matings of the new animals, that occasional fertile matings to chaus and libyca occurred.

Perhaps even looking for a clear start is too optimistic for many hybrids have been reported over the years. Some early records are questionable, but certainly hybrids between the domestic and a number of the true wild cats have occurred over the centuries. It is currently very hard to draw a line with the recovery of F. silvestris in Britain between 'true' silvestris and silvestris/domestic crosses, especially when dealing with a number of generations removed from a cross. In remote corners, where domestic cats can have been feral for generations, meetings and matings with silvestris become probable rather than possible.

Although the dice seem heavily loaded in favour of F. silvestris/libyca as progenitor of the modern domestic cat, doubts have often been raised on the basis of the extreme intractability of silvestris. Certainly on rare occasions the cat has been tamed and lived when young in domestic circumstances. However the newsworthy nature of such an event is a reflection of how unusual this is. Even as young kittens the European wildcat can seem like fury incarnate – a truly wild animal.

On the contrary the Jungle Cat, Felis chaus, behaves in many ways like the domestic cat. It tames fairly well and although India may be noted for its large dog rather than cat population I have found in the more forested areas of Northern India and Southern Nepal that the Jungle Cat can be seen hanging around habitations, being fed scraps and on occasions being made into a pet.

It is thought not to be a major contributor to the domestic line as the museum animals measured are much larger than the other contender. However the other contender, silvestris/libyca, is not a true match either. I have seen in the wild, mature Jungle Cats comparable to domestic cats. It is apparent that certainly chaus and catus have a wide range of sizes when mature and probably silvestris too. I am inclined to think that F. chaus played a bigger role in initial crosses or subsequent crosses in the development of the domestic cat than is currently believed.

Whatever their origins, it seems certain that all domestic cats range of coat colours do not need exotic hybridising to explain them, the normal genetic events and normal low level gene mutation since domestication can account for these.

Despite this, many feel that the exotically long hair of the Persian Cat derives from a cross with the central Asian Pallas Cat. Similarly as Siamese cats show their own character, voice and coat, it has been suggested these may claim at least part descendancy from the Golden Cat of SE Asia. As fertile crosses between wild and domestic cats occur, it is hard to see how this can not have been happening between free living feral cats and probably most of the world's small wild cats for as long as close contact has been maintained. The Siamese cat, like the ancient Egyptian cat, enjoyed a sacred status and had a restricted ownership of aristocracy and priests, but unlike in Europe seems not to have suffered a fall.

The cat in Egypt

Nothing connected with cats is apparent in Egypt before the 5th dynasty and the later 12th dynasty paintings of around 1900BC show cats hunting. By 1250BC however, they can be seen in more obviously domestic settings.

Despite the hunting prowess of the cat, house-snakes are believed to have been the ancient Egyptian rodent controller. The ever alert fierce but tamable mongoose in neighbouring Sumer was considered better for rat-catching:

'A cat for its thoughts! A mongoose for its deeds.'

This would also seem so for Egypt, particularly in the absence of poultry, as mongooses are particularly fond of them. It is likely that scavenging led the cat near man more than the number of mice at granaries when present day free-living cats are considered. By the time that cats were domesticated properly, the Egyptians were worshipping the cat-headed goddess Bastet, based particularly at Bubastis.

It is assumed today that if there were a cross in the wild producing the ancestor of our domestic cat which then exploited a scavengable food source of Egyptian towns (rather than being initially bred in captivity) that such an animal no longer exists in the wild state. However, today there is a flux between feral and domestic animals, as initially there would have been between wild and domestic. Whenever a modern householder adopts a modern stray or tractable feral animal an echo is being played out of those first encounters. The feral cat of today is following the role of his first ancestors and in parts of the Middle East it is just possible that some of the fiercest most intractable feral animals living away from man may have a strong undomestic line to the first undomesticated Felis catus.

Cats are thought to have been venerated in ancient Egyptian households, to the extent that on the death of a cat its owner shaved off his eyebrows to show remorse. To kill a cat was unthinkable. It was so awful that in the mythology such actions were among the worst blasphemies an invading army could do. Although the deities Seth and Horos were once gods representing the two lands that formed Egypt, they became seen as Horos representing Egypt and Seth as any invader capable of any crime including desiring to eat the sacred cat in the presence of its mother, Bastet.

Egyptians were loath to let cats out of the country and are believed to have actively sought out any that were beyond the borders. This has led to the suggestions that domestic cats may have 'turned up' first elsewhere but had been hi-jacked.

Despite the Egyptian possessiveness over their cats, with Roman Imperial contact they gradually found their way into the European Roman Empire as rare and exotic household pets.

It seems that the name 'cat' followed the animal from north Africa. There are a number of forms in the Berber language. The Latin form 'cattus' is first found used by the fourth century Palladius derived probably from the berber or semitic, when Palladius recommends the use of cats for removing moles from artichoke beds!

Is the cat domesticated?

Is the cat a domestic animal? This is a difficult question for the cat co-habits with man and yet continues to live in a free way while doing so. The high proportion of cats living a feral or stray existence demonstrates the cat's ability to fend for itself. Further, the cat was the last common animal domesticated, long after that gullible pack animal, the dog, and later even than the llama and the goose.

Due to its elusive status of domestication the cat more than any other animal forces a closer examination of the term domestication. It takes us to the world of 'chicken and egg' questions. Did we 'domesticate' animals in the first place, or were there certain species of animals that already had certain characteristics of tractability such as rapid habituation to strange circumstances that enables us to 'take them under our wing'? We may have bred for docility since, but nonetheless the latter seems more likely. Herd animals today such as sheep and cattle receive very little direct affection from man, associations being based on normality (the rest of the herd does not run when a man approaches so neither does the lamb or calf). It seems that these are just very tractable animals that we have confined rather than magically domesticated.

Cats however are hardly ever confined or owned in quite the same way. This is even recognised in British Law by not expecting it to behave as a domestic animal. (Suggesting a licence for cats therefore requires careful consideration of the cat's legal status.) They form territorial and home range attachments to areas and a modified group attachment. The cat is still only relatively domesticated and is capable of remaining uncompromisingly independent. It may be this inherent self-reliance that allows the cat to seem aloof to some.

The cat reaches Britain

As bone remains of cats (particularly of the body) are not instantly distinguishable between wild and domestic, the early British bones have been sorted mainly by association. The cats found at an Iron Age site at Glastonbury in Somerset are believed to be wild as they were found along with other wild carnivores of fox, otter, weasel, marten and polecat. In contrast to this cat footprints found on clay tiles that had been left to dry in the Roman town of Silchester (Berkshire) due to their location, are unlikely to be wild. Certainly the Imperial Roman world had adopted cats and aided their spread and the Romans may have brought the domestic cat to Britain. From well into the British Roman period (4AD) a cat has been found that died in a house fire in which the skull is unambiguously domesticated. The villa belonged to a wealthy man and it is likely that in the early years they may have been only associated with the wealthy as 'rare and exotic animals from the East, prized by ancient kings'. Although it is likely that the Romans brought the cat to Britain suggestion has been made that this was first undertaken by Phoenician traders.

Following the withdrawal of the legions back to Rome, England was in some turmoil between warring factions of invading northern Europeans, Saxons and Danes, nonetheless Christianity took firm root in the seventh century AD. Christianity fought a rearguard action against the influence of the earlier gods, even of pre-Roman echoes, and it is possible that this became formalised and hardened into the later obsession with devils and witchcraft.

Certainly the initial reception of the cat in Britain was favourable and remained so for some centuries. The first written reference to domestic cats in Britain was the passing of a protection law by a prince in South Wales, Hywell Dda, the Good, in 936AD. This law putting a high reparation value in wheat on not only adult cats but also on kittens with and then without eyes open (in descending value).

As Christianity caught hold so did the cat, and it is noteworthy that cats were the only pets allowed in monasteries and nunneries. By medieval times their numbers had built up to some extent for the only fur permissible for the trimming of clerical dress was that of the cat. Perhaps the functional outlook of the times encouraged the keeping of cats as pets, to deter rodents, and at their demise to provide much needed warm linings to garments. Life was then more proscribed and law and custom even insisted on identifiable clothes and particular furs to be worn on the basis of status and function. Archbishop Corboyles 'Cannons' (1127) ordained 'that no abbess or nun use more costly apparel than such as is made of lamb's or cat's skins. Edward III's laws restricting the users of apparel specify cat skin. The description of a domestic cat in '*De Rerum Natura*' (Bartholomew Glanvil, 1398) is an affectionate and accurate view of a cat:

'The catte is a beaste of uncertain heare and colour, for some catte is white, some rede, some black, some spewed and speckled in the fete and in the face and in the eares. And he is . . . in youth swyfte, plyante and merry and lepeth and reseth on all thynge that is before him; and is led by a strawe and playeth therwith. And is a right hevy beast in age, and ful slepy, and lieth slily in wait for myce . . . and when he taketh a mous he playeth therwith, and eateth him after the play . . . and he maketh a rutheful noyse and gustful when one proffereth to fyghte with another.'

In this not only are we told of the playful domestic nature of the cat, but are clearly shown the wide genetic variation in coat colours by this date.

At the same time in the late 1300s comes the story of Dick Whittington, thrice Lord Mayor of London and his rare and luck-bringing beast – the cat. Cats were then still favoured but nonetheless it seems the cat that Richard Whittington amassed his fortune from on the way to London were robust ships called 'cats' taking coals from Newcastle to London.

The Hunter Hunted

The medieval landscape was in part administered for hunting, particularly in those large tracts of land established as Royal Forests. Certainly cats were among the beasts of the chase mentioned in royal afforestation grants.

When domestic cats were rare the animal chased would have been the wild cat, but as the numbers of domestic cats built up and in turn many turned to living a feral existence they too would be hunted (particularly with the tendency for rural feral cats to revert to wildtype markings and cross-breed with the wild cats).

As now the domestic cat is such a part of the everyday household a number of people find it unpalatable to believe that the cat was ever hunted in 'the chase' and allude to the 'other' chase cats, such as the martin-cat. However it is quite apparent that both animals were pursued.

The medieval chase term of a 'clowder of cats', referring to a group of cats, is far more likely to refer to stray or feral cats than to wild cats.

The deity becomes diabolical

European society had undergone a period of stablizing consolidation from small warring tribes into larger territorially established kingdoms and Christianity was really only consolidated over Britain and Europe a little over a thousand years ago. The secular side of the church became powerful and even when there was only one Pope at a time, Rome to many seemed to have become another princely state. Dissension led to the early rumblings of the Reformation, and established thought felt itself threatened.

In countering the new protesting thoughts in religion and science the Church found a unifying force in rooting out adherents to the earlier pre-Christian folk religion. This was fully absorbed as the enemy within of the devil and his supporters persecuted as heretics, witches and wizards. Opposition was sublimated as evil and therefore of the devil. The seeking out of witches lasted for three centuries, spanning the period of the violent separation of the Catholic and Protestant churches. Curiously the new formed Protestant sects in excessive zeal of puritan self doubt, sought the same rallying ruthless hounding of witches.

These were violent times when many good men and women were burnt due to religious conscience, as well as many defenceless women as witches. The fate of the cat suffered a terrible turn of fortune during this period. From being a deity, then monk's companion, it had become a witches' familiar. As is clear from 'De Rerum Natura' the character of the cat was no different then from now, but man's concept of his world and his role in it has fluctuated over the years and with it his attitude to the treatment of cats. The cat more than any other domestic animal has suffered the swings of man's theological quest. Whether devil or deity – certainly the cat has been seen as supernatural. The reason for this would seem to lie partly in the detached independence of the cat, which even in play seems uncanny. Partly it is also a reflection of the cat's anatomy and behaviour. The cat population had apparently grown to the point where then as now any lonely elderly lady could keep a cat as companion.

I live in a village only a few miles from Chelmsford in Essex, the town in which more witches were hanged than anywhere else in England. It is not by chance that Essex has also the strongest tradition of Protestant sects in England. The emergent sects often saw 'Popery', Satanism and Sorcery as having a common aim in driving mankind to damnation. Doubtless many witch trials picked on deluded solitary people, but perhaps there were some practising the old spirit worship, for it is strange that the witches arrested came mainly from the areas first settled in the county. Either way a peculiar feature of British witches was the association with familiars.

The first trial specifically for witchcraft was in Elizabeth I's reign in 1566, appropriately at Chelmsford. From it can be seen the believed role of the cat in witchcraft through this traumatic period. Unfortunately the fairly normal action of cats being seen to leap through a window of the cottage where they lived at twilight or to open a latch on a door was considered by some magistrates, during the witch hunt era, sure evidence of the animal's being a demon.

1566: The first witch trial and Sathan the cat

This trial was crucial with the Queen's own attorney trying the defendants. These were Elizabeth Francis, Agnes or 'Mother' Waterhouse and her daughter Joan. These ladies were believed linked in witchcraft via a cat that was:

'a whyte spotted catte . . . (and they) . . . feed the sayde catte with breade and milkye . . . and call it by the name of Sathan'.

This trial and the execution of Agnes Waterhouse unleashed the original witch-hunt and in Essex over only the next twenty or so years some 150 people were taken before magistrates for witchcraft. In most, a telling feature was believed involvement of familiars, the evil spirits that acted as agents of the witches, animals that seemed to be able to change form. The animal was not always a cat, but they were so often involved that people became scared of even talking in front of a cat. The idea that cats have nine lives seems to have arisen from a statement in Baldwin's 'Beware the Cat' (published about 1560) that: 'it was permitted for a witch to take her cattes body nine times'.

A theological change had transformed both European and protestant early American settlers' outlook on the cat. The divinity had become a demon or even the devil himself.

It was a long purge, the last official execution for witchcraft took place in 1684.

Fiery torment for devils

The cat had been so devalued and so associated with the devil that cats were burnt in wicker cages all across Europe. Sacrifice of living animals to the flames had been an old practice of the ancient celts particularly of ancient Gaul at high festivals and now domestic cats were burnt at large festivals. They could not suffer enough for the crowds, who saw in their wailings the devil getting a taste of his own medicine. To prolong the agony the baskets containing the cats were hung initially well above the flames from a tall pole.

During the seventeenth century the Protestant world was becoming more assured, less under threat and the Age of Reason was ushered in. However, even under the scrutiny of the reasonable Royal Society, devils, witches and their trappings did not instantly dissolve. Divine miracles were thought to have ceased by the end of the Apostolic Age but perversely a few philosophers such as Henry More and Joseph Glanville (author of '*Philosophical Considerations Touching Witches and Witchcraft*' 1666) made the case that to abandon belief in witches meant accepting atheism. They saw witchcraft as the only tangible proof of supernatural activity – howbeit for the opposition.

By the early eighteenth century official action against witches had ceased, as belief in them waned. Nonetheless in rural areas, distant from centres of modern thought beliefs take longer to change. In 1718 William Montgomery of Caithness one night could no longer stand the yowlings of courting, sparring cats outside his house so he descended on them with an axe and some diehards had their beliefs that witches could turn into cats confirmed when the light of day showed that two old women of the district had died overnight and another had a bad cut in her leg.

Gradually as reason continued to pervade and theological beliefs continued to wane, the cat was able to lose the associations of spiritual life, whether divinity or demon, and settle into the role of pet and household companion.

Naturally as an enlightened man, a leader of thought – and also a kindly man, despite his show of irascible blustering, Dr Samuel Johnson had a household cat as Boswell related: 'I shall never forget the indulgence with which he treated Hodge, his cat; for whom he used himself to go out and buy oysters, lest the servants having the trouble should take a dislike to the poor creature. I recollect him scrambling up Dr Johnson's breast, apparently with much satisfaction, while my friend, smiling and half-whistling, rubbed down his back and pulled him by the tail; and when I observed that he was a fine cat, saying 'Why, yes, Sir, but I have had cats whom I like better than this,' and then as if perceiving Hodge to be out of countenance adding 'but he is a very fine cat, a very fine cat indeed.'

Cats were no longer rare and exotic pets. They had built up in numbers and as so many lived feral or semi-feral, scavenging in the growing urban sprawl. Household refuse thrown daily into the streets would only be scraped to huge mounds after it had lain for some days. Rats and mice were plentiful. Paradise for an independent living cat. They had need of their supposed nine lives. The stain of black magic demons lingered and small boys needed little excuse to be unspeakably cruel.

During the eighteenth century cats were often the object of cruel entertainment, with, it seems, little pity spared for them by many still haunted by the echo of witchcraft. One such was on a high day or holiday to suspend a barrel of soot containing a cat, and the men of the town would run or ride under the barrel striking it with clubs trying not to be under it when it eventually broke.

It was often the custom to put the body of a cat, often with a rat, into a wall or thatch of a new house during the seventeenth, eighteenth, and even nineteenth centuries as late as the 1880s. This talisman was meant to deter both rats and the devil from entering the house.

Today we tend to assume that cats have 'always' been *popular* household pets but antipathy and dislike of cats was still predominant during the Victorian period. A Popular Encyclopedia in the 1870s commented that the common opinion of the cat was that 'her treacherous calmness of disposition needs but slight provocation to be changed to vengeful malignity'.

Writing in 1889 Harrison Weir noted the cat had suffered:

'Long ages of neglect, ill-treatment and absolute cruelty, with little or no gentleness, kindness. . . . Should a new order of things arise and it is nurtured, petted, cosseted, talked to, noticed, and trained with mellowed firmness and tender gentleness, then in but a few generations much evil that bygone cruelty has stamped into its wretched existence will disappear. . . .'

It was Harrison Weir who:

'. . . conceived the idea that it would be well to hold 'Cat Shows' so that the different breeds, colours, markings etc. might be more carefully attended to and the domestic cat, sitting in front of the fire would then possess a beauty and an attractiveness to its owner unobserved and unknown because uncultivated hertofore'.

He instigated and judged the first Cat Show at the Crystal Palace on 16 July 1871, having first decided the number of classes and 'points of excellence' for judging the various classes. He hoped by this to improve both the appearance and thereby the lot of cats for, as he later wrote: '. . . as little or no attempts have been made, as far as I know of, until cat shows were instituted, to improve any particular breed either in form or colour'.

The lot of cats has indeed since improved, but Harrison Weir was so disgusted by the sudden appearance of those whose 'Principal idea . . . consisted not so much of promoting the welfare of the cat as of winning prizes' that he resigned as President of the newly formed National Cat Club.

Certainly today the continued inbreeding of certain lines for show purposes, such as some snub-nosed breeds, is not in the best interests of those animals.

The cat and the gamekeeper

The Victorian era also saw the development and flowering of the great shooting estates, so the days of cat as hunted animal were not over. With misguided zeal gamekeepers shot or trapped anything that could possibly be a predator – including unfortunately man on occasions when poaching. The feral cat population of Britain, although less dense in rural than urban areas was considerable, and rural feral cats are efficient predators. However, as for other predators and rigid control by gamekeepers, the cats would not have been feeding anything like exclusively on game birds and the removal of large predators increased the number of brown rats.

Nonetheless between 1874 and 1902 along with nearly a hundred polecats and thirteen pine martens, some 2310 cats were shot on just one group of Welsh estates.

The First World War saw the demise of most estates, but today the active control of feral cats continues unabated (see chapter 14).

The cat today

The association of cat and man has not been a quiet one and good or bad, deity or demon, the cat has been seen as having supernatural powers. Certainly the cat can perform some extraordinary feats and as *the* mammalian carnivore has time to appear contemplative, independent and apparently not in need of human support.

As our outlook on to the cat has depended on the beliefs of the time, it is small wonder that to a lesser extent the same occurs today, when the cat is firmly assimilated as a house pet. There are pro-cat people and anti-cat people. Very often anti-cat people are fiercely pro-dog.

Once again it is the outlook of the person involved, conditioned by their upbringing, that causes the difference. Dogs are pack animals having evolved group hunting in open plains. Their 'peck-order' demands that the owner establishes himself in a 'master' role and the house dog will greet him with submissive postures, ('fawning' behaviour in the eyes of a 'cat' person).

The cat, on the other hand, although having greater social structure than once thought, remains a solitary hunter, of woods and undergrowth (viewed as 'aloof' by the 'dog' person).

However, throughout its history the cat remains the same, not aloof, not fawning, just gloriously self-sufficient.

Nonetheless to the biologist there are not sides to be taken, rather different strategies that animals have taken in exploiting available food sources to ensure survival. From that outlook domestication becomes just a survival ploy embarked on by the animals involved, and less of a firm distinction.

CHAPTER 1

A Cat of Many Colours

Cat and man may have lived 'paw in pocket' for a long time, but cats have maintained an independence which man seems to have curiously respected. Most domestic animals have had a direct economic attraction for man, which has resulted in his selectively breeding certain characteristics. This has largely not been so for the cat. This could be attributed more to the inherent free nature of the cat, with the matings of cats being harder to restrict than matings of most domestic (or perhaps *more* domestic) stock. There had been little or no development of breeding lines in cats before the first cat show to be held at the Crystal Palace in 1871.

Apart from today's tight confines of the cat breeding world, cat genetic patterns, usually seen from coat colours, are thought to be from mainly random breeding and largely free from selection. However, man has been associated with cat distribution patterns, both directly and indirectly. The majority of mutations in the cat show differential selective sensitivity between town and country.

While most wild mammals display agouti (or brindled) colouring deriving from a banding on each hair, most domestic animals are non-agouti (uniformly pigmented hair). The cat however is an exception being prominent in both forms. Non-agouti colouring is favoured by the urban habitat and is commonly black.

The black and yellow/grey striping of individual hairs in the agouti coat are also the basis for that notable cat marking – the tabby. Areas of the coat lose the pale part of the agouti hair (as in the totally non-agouti cats above) and so seem to have black stripes set in a brindle coat which we see as the tabby stripes. The 'wild-type' or mackerel pattern is similar to that seen in the wild cat of Scotland (Felis sylvestris). However when the great Swedish taxonomist Carl Linnaeus (who originated biological classification) designated the domestic cat as Felis catus he used the blotched tabby. Retrospectively it would have been better to have opted for the mackerel type, the blotched tabby being a recent mutant. Nonetheless, the blotched tabby is the 'typical cat'.

The blotched tabby is believed to have originated in Britain and is fast spreading throughout the world. The mutation may have appeared during the reign of Elizabeth I, coinciding with Britain's rise in seapower. The spread of the gene in its new form was assisted by British colonization. As numbers rapidly increased in the original British population, each subsequent wave of settlers took a higher percentage of the new form among their cats to the fresh distant lands. It would seem quite proper then to claim the title of 'British Imperial Cat' for the apparently humble blotched tabby! The blotched tabby seems to have some advantage over the wild-type, mackerel tabby. In the truly feral cat population of Tasmania, the other side of the world from Britain, it is 'winning hands down'. It is hard to pinpoint the source of the advantage. In Britain I have found in the crowded conditions of suburban domestic populations a few large unneutered blotched tabby toms to assume the role of cunning aggressive warlords. In rural conditions I have found them to be efficient hunters. It may be possible that these are associated traits that just give sufficient edge to the 'new' blotched tabby.

When the non-agouti colouring is not restricted to stripes, but is all over the cat, then this is most commonly seen as a black cat, but despite being black it still carries the tabby alleles. Thus the black cat can be thought of as a masked undercover agent of the tabbies. Indeed the undercover nature can sometimes be detected in black cats when tabby lines are glimpsed in a certain light. As the blotched tabby started in Britain, so the black cat mutant started in Phoenicia, and that ancient seafaring nation assisted its spread.

The cat about town

Whatever its origins, the black cat is common in London's feral population, especially in the West End where a high proportion of

London's long established cat colonies are found. In this area however the non-agouti black phenotype is accompanied by piebald (white) spotting. With the black cats this gives the typical appearance of the 'West End' or 'Clubland Cat' as a dapper puss-about-town in dinner jacket and white gloves, (indeed 'Bustopher Jones wears white spats', as T. S. Eliot notes!)

In London's Fitzroy Square, near the Post Office Tower where I have particularly followed the behaviour of the cats, (see Appendix) the range of proportion of white on the cats can be seen to encompass most of the possibilities. The matriarch of this particular group had been nearly all white. Nonetheless the West End cats are mainly white in shirtfront, paws and perhaps nose (and in terms of their alleles can be thought of as heterozygotes for white spotting. The matriarch referred to is on the other hand a homozygote for white spotting).

An examination of 700 cats that had been destroyed in London at animal welfare clinics found some degree of white spotting very common (particularly the heterozygote state). Black was very common as was the blotched tabby (both having allele frequencies well above the wild type mackerel pattern).

Matings were believed to have been largely random from the sample used which was partly domestic and partly free living.

In the stable West End 'Clubland' cat colonies totally random matings may not occur. This is not because of human interference, but territorial controls. Toms will visit neighbouring colonies, two or more blocks away, but such is the size of the area containing these West End groups that similar genetic meetings occur. Within peripheral groups near Covent Garden this localised hold slackens as does this type of colony. Food supply in the West End area has not been a limiting factor for cats due to extensive auxiliary feeding and scavenging. The habitat for these successful colonies is high in railings and basement areas, parked cars and strong overhead street lighting. In such conditions of sharp shadows it seems that the black and black and white cat has a survival advantage and thrives.

CHAPTER 2
Design for a Hunter

The skeleton of the cat is the epitome of hunting design.

The claws make the initial impact on the prey and to protect the base of the claw, the bone that the claw lies on has a well developed protective hood.

We usually think of cats retracting their claws, but this is to misunderstand not only their behaviour but also their anatomy. In

A skeleton built for speed – when required!

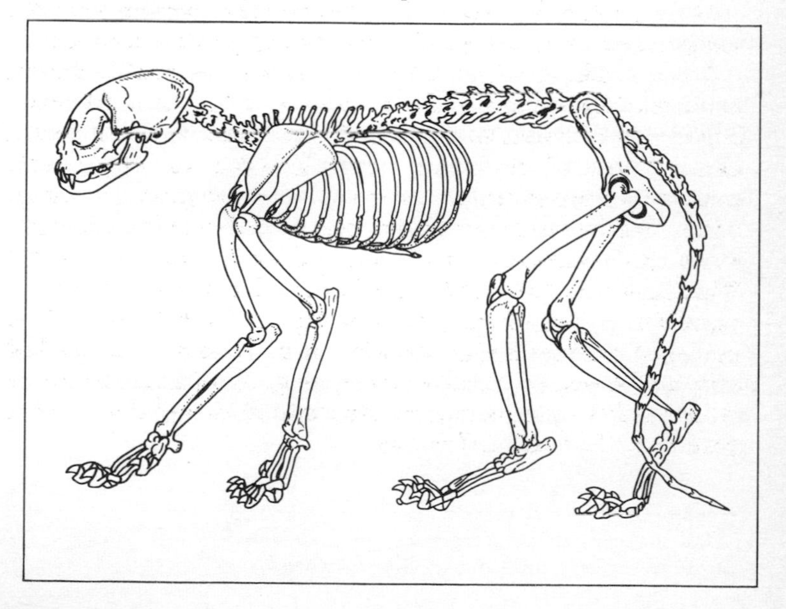

the relaxed state the claws are 'sheathed' and when required to strike they are actively *protracted*. The paw digits are normally spread 'hand-like' at the same time. When relaxed the claw supporting end bone is held against the next one by a retractor ligament, but to protract the claw a flexor muscle actively contracts and pulls a tendon under the foot straightening the bones ready to strike.

Cats are digitigrade, that is, they walk on their toes and so, when running, gain speed over plantigrade animals like us. The plantigrade animal has settled for economy and stability, while the hunter demands speed. However this makes the 'ankle' joint, the main mid-leg joint, look as if it is a backward pointing knee.

The cat when hunting does not go in for long chases, but it will often incorporate short fast sprints in its stalk approach and final explosion. When running fast the retarding action of the forelimbs is almost completely lost, for the cat fully extends the forelimbs and arcs them downwards and backwards *before* contact with the ground. This with the arching of the animal's spine reduces the check on the backlegs from the front ones. The very flexibility of the spine that enables the cat to wash over most of its body beyond the dreams of a contortionist, also lets the cat increase its stride length while running.

Cats have more rounded chest cavities than us and have shoulder blades (scapulae) at their sides rather than their back, so 'in-line' walking, essential to a climber, is easier. By also having only a vestigial collarbone (or clavicle), where really only long ligaments connect the sternum with the shoulder blades, the scapulae are freer to move and again increase the running stride.

Red in tooth

Cat teeth are made for meat eating and fit into the classic meat eating skull. Cats have savage looking canine teeth to grip and kill the prey and the most specialised long sabre-like carnassial teeth

---

The tail angle balances the cat in leaping to the top of a wall, holding position, swinging round and walking forward.

Wall Walking

along the jaws for slicing through flesh. The jaws are short to increase power and the skull's design incorporates strengthened bony arches to cope with the tension load in the jaw muscles.

Vets often find themselves presented with cats with fish bones stuck on or between teeth, resulting both from meals fed at home and from scavenging. Long sharp bones become jammed between teeth, frequently breaking leaving a part working into the gum.

Inspection of the teeth provides a guide as to age grouping of a cat. In both domestic and feral populations, as the cat ages through its mid-years, the gums recede leaving a clear ring of their original height at the wide point of the tooth, the tooth then narrowing rather than widening into the gum. Similarly the canine teeth develop longitudinal grooves that darken with advancing years.

From examining the teeth of ninety-two adult feral cats taken from a number of colonies I found their teeth to be in remarkably good shape. Any redness or swelling of gums was not any more frequent than in the domestic population. Only 5 per cent had appreciable quantity of scale deposited on their teeth and only two animals had broken teeth, one a male of almost 9lbs and one a female of a little over 6lbs. Seven animals had some teeth missing, one of which was a very large tom of nearly 17lbs weight which had a number of teeth missing. Thus some 10 per cent of the cats had either broken or missing teeth.

While in most of the different sites cats had very good teeth, all of the cats with damaged teeth came from two urban East London sites rather than rural sites, and most of those from one site. It is also worth noting that most of the cats with scaled teeth also came from this same site. The diet of these particular animals consisted of scavenged material heavily augmented with poor quality sloppy food. Most auxiliary feeders feed either with proprietary cat foods, usually complemented to add variety with fish, liver or chicken, and milk. This was not the case for this colony, where the staple item of diet was larded white bread. The animals consumed large volumes of this material, giving them very thin-walled distended stomachs. The stomach lining when looked at under the Scanning Electron Microscope was also found to have suffered damage apparently from this regime.

Lap before you eat

The tongue in many mammals has a fairly bumpy top surface, but in the cat this has developed into a particularly useful tool. The bumps have become cornified backward-pointing spines, each shaped like a smaller version of the tongue. As these spines point towards the back of the throat they usefully serve as a rasp, breaking off and gripping pieces of food. In the domestic cat that often lives on soft tinned foods the gripping function of these papillae plays a greater role in picking up food than in the wild state.

In the fastidious cat these flexible hard spines are most frequently applied to grooming. That the cat's tongue is easily as efficient as a fairly expensive brush can be seen from the mat of hairs left on your armchair. We usually say 'Oh the cat's moulting' or something slightly stronger, forgetting that these hairs did not spontaneously leap off the cat, but that in the comfort of the chair the cat has carefully and methodically brushed them out with her tongue.

Taking great care of its coat can result in the cat swallowing some most unpleasant material. I was once painting bitumen onto an outhouse roof when my cat leapt up to see what was going on, landed on the fresh black bitumen and skidded straight off the other side with balls of thick black tarry goo on her paws. She must have ingested a fair amount as she hid and washed it off, but fortunately was all right.

Feral and house cats, in their normal cleaning regime must ingest a lot of lead from particulate lead from vehicle exhausts settling amongst road dust. The intake of leaded dust by cats when washing must, despite the cat's tongue brushing a large amount off the coat without ingesting, be considerable.

In domestic life the cat's tongue fulfils another feeding function it does not meet in the wild – lapping up milk. The ability of the cat to semi-cup its tongue helps but it is the large number of papillae which hold a relatively large amount of milk on the tongue due to the involvement of surface tension. In the wild this elaborate system is normally used to pick up water from puddles and pools.

In lapping in a shallow dish the forward motion of the tongue will of necessity, propel liquid forward and so a large amount is missed. This does not matter in a pond, but in a saucer the cat often tries to increase the amount it laps and so will lean across the saucer and lap up the incline. The usual result is that a fair proportion is thrown out onto the carpet. Don't blame the cat however, just choose a different dish, preferably saucer-shaped but with the edge vertically raised for about an inch or so and this instantly catches the excess and also the cat drinks a little further back from the edge. However, cats will not willingly feed from anything set too deep and bowl-like probably as they are unable then to keep an eye on the world.

It seems that unexpectedly, in drinking from a volume of liquid deeper than that normally encountered in saucer and puddle, but found in pond or stream, then the tongue folds and the back acts like a ladle, scooping in reasonable amounts (although still supplemented by that adhering to the front).

Fur is marvellous stuff – particularly for cats! Not only does it conserve body heat – but it is also equipped with a self-cleaning, self-aligning mechanism. This might at first sight seem almost superfluous for the fur-conscious meticulous cat that always seems to be washing, but the mechanism also works in close conjunction with licking!

Each hair is covered with overlapping scales (only 0.0005mm thick), that like roof tiles have an uncovered free end pointing in the same direction, on hair towards the tip. The free edges of the scales work like teeth in a ratchet. If one hair falls or is pushed out of alignment low over others, then any movement of the animal will cause the stray hair to click up the shafts of the other hairs, only in one direction, until it falls back into alignment. One swift rake with the papillae of the tongue will speed the process up. Bits caught in the fur will similarly be worked to the surface on the ratchet-edged hair scales, again a process accelerated by the raking tongue papillae.

Cats eyes

Cat the hunter and predator can be recognised from its forward facing eyes which give better depth of field for pursuit. Herbivorous potential prey such as rabbits have all-round vision by having eyes in the sides of their head to see a predator coming from any direction (this makes stalking rabbits harder for cats).

The domestic cat's eyes have adaptations to a nocturnal way of life. Like other nocturnal hunters such as the owl, cats' eyes are very large in relation to their skull size when compared to daytime animals. This has allowed a huge field of vision for each eye of around 205°. It also lets the cat's eye take in 50 per cent more light than ours.

Although cats can see colours their eyes do not receive colour anything like as well as our own eyes. This does not mean that cats have inferior eyes to ours, just that their life demands different requirements of their eyes. There are two types of light-receiving cells in the eye's retina – rods and cones. The cones discriminate colour, but are less sensitive at low light levels than rods. Consequently to make the most efficient use of any available light at night cats have a high proportion of rods – around twenty rods for every cone – compared with about four to one for man. Their colour vision is likely to be proportionately less than ours during daylight hours, but they can see far more than us after dusk.

Another trick that improves their night vision is seen everytime a cat looks back into the glare of car headlights with brilliantly reflective eyes. So well known is this that the reflectors that mark the crown of the road are also referred to as 'cats-eyes'. It is perhaps less widely realised that this is not just a quirk but a functional adaptation to night life and that your car headlights are being reflected back from biological mirrors. Behind the rods and cones is a coating of light-reflecting crystals, the tapetum, so that the cat makes use of all available light. Our retinas that do not have this do not absorb all of the light that enters the eye, but in the cat any light straying through the retina is immediately reflected back onto the sensory cells. At the level of these cells this double source of the light is bound to give some 'ghosting' as multiple images which should reduce the clarity. However, by these adaptations to a night life the cats seem to be able to discriminate at a sixth of the light level we require. Nonetheless what should not be overlooked is that the optical clarity of our eyes is not as high as the view our brain visualises, for our neurones clean out visual 'noise' and it may well be that the cat similarly can 'sharpen up' its picture.

The wildcat with its relations, including the domestic cat, has eyes contracting into vertical slits, unlike the large daytime hunting cats such as the lion where the pupils stay circular. The cat's eye is so sensitive to light that it needs a greater degree of fine control than us over the amount of daylight it lets in otherwise it would be dazzled. This seems to be managed, not only in the cat but in other nocturnal or primarily low-light animals such as the geko or some sharks, by having an iris or shutter arrangement that instead of coming to a smaller diameter hole closes as a vertical slit. This has always given the cat a mystique over most regularly encountered animals.

In practical terms the cat can at night have very wide circular pupils to allow in most light. We only regularly achieved quite the same extent by Renaissance women dropping atropine into their eyes. Shutting to a slit during the day allows fine control while only allowing in a small amount of light. To act further against dazzle, rods do not function so efficiently as cones at high light intensity, and rather than being a disadvantage this is another advantage in having less colour vision for a nocturnal animal with its image intensifiers, so that it is not swamped.

People are sometimes startled when they see for the first time their sleeping cat with its eyes open and they do not see pupils but just white voids. However if a shadow passes over the eye the pupil reappears as the 'third eyelid' or nictitating membrane is horizontally whisked away. This membrane is translucent and by closing it over the eye the cat can reduce its sensory imputs and have a daytime doze, while still remaining alert to shadowy approaches.

Sound sense

It seems that cats have far sharper discrimination in hearing than dogs, which accords with their role as the nocturnal lone hunter.

Research has shown that cats can distinguish better than either dog or man between two sound sources that are close together. Cats can also detect sounds that are at different heights more accurately than dogs, but unlike dogs cats are used to hunting after mice and voles and climbing in trees after birds.

Cats have also shown, in research conditions, the ability to distinguish sound depth better than dogs. When a sound is made further on but in the same direction as another sound, cats are always able to follow the direction, while dogs are only able to detect the first sound. The difference would seem to allow the cat to locate its prey exactly even in the dark for an accurate pounce, while the pack dog will run towards the prey by day and then rely on sight and group tactics.

Detection of small mammals hidden by long grass depends heavily on the cat's ears. The pack hunting ancestors of dogs on the other hand aimed at pulling down larger prey.

The upper frequency limit that the domestic cat's ear is physiologically capable of responding to is around 100KHz. However, above 35–65KHz sensitivity is low, but certainly cats have been trained to respond to tones of up to 60KHz. These levels are well above anything the human ear can pick up with its upper normal limit being between 15–20KHz. Although dogs are known to respond to high frequency whistles they have a physiological capacity of 60KHz and with a practical upper limit of between 15–35KHz are not even in the same league as the cat. The clue again lies with the prey species.

Generally, unlike for man or dog, the small prey species of the domestic cat, whether bird or small rodent, make high pitched cheeps, squeaks and rustles and so the high pitch sensitivity can have a survival advantage – particularly to the free-living cat.

Nonetheless, as with people, the cat's sensitivity to high notes fades with age and a cat of four and a half years does not have the same wide range as a one-year-old.

My house-cat's sensitivity to high pitched sounds is clearly demonstrated by her reaction to my whistling. She will invariably leave another lap and come onto mine, purring and with a glazed look in her eyes, but then proceeds to rub her head very definitely against my mouth effectively stopping the sound. This is particularly so in response to the higher notes and louder sounds.

Strangely we have a low hearing threshold and many animals including the cat do not have acute hearing below 200–300Hz. Even so, the cat has a range of well over ten octaves to our eight and a half, but then the domestic cat does have 25 per cent more nerve fibres than we do in its auditory nerve.

The large cones that are the cat's external ears or Pinna are well equipped with over twenty muscles, allowing the independent swivelling of an ear in the direction of a sound, to front, side, or to an extent backwards. Sound location ability is lessened with directional swivels while moving, consequently the initial ear sweeps would seem to be made with the cat stock-still.

The well-muscled ears figure prominently in the cat's range of facial expressions, particularly in aggression where the ears fold at the back and are laid flat on the head.

The tube from the Pinna to the eardrum is well lined with ceruminous (wax) glands and well dotted with protective hairs. The amazing sense of balance of the cat is largely attributable to part of the ear complex, a series of fluid filled canals containing sensory hairs. The relative movement of the cat to the fluid bends to the hairs and gives an awareness of movement of position. Contrary to some cat owners I have found that small rapid movements of fluid in the canals encountered when travelling by car can unfortunately bring on motion-sickness in the cat.

The cat's whiskers!

The very phrase 'the cat's whiskers' carries with it a meaning of something special. Indeed for a cat they are something special, for being primarily a dusk and night hunter the alert cat needs a high sensory input from particularly sound and touch to aid its visual cues. They are of great help in dim lighting conditions and in moving among branches in trees or in undergrowth, but they are no less important to the urban cat in negotiating gaps in fences. The importance of whiskers is seen in a cat that for whatever reason has lost them; it is slower and clumsier in finding its way about.

Vibrissae (or whiskers) are larger and stiffer hairs that are well supplied with sensory nerve endings at their roots, ready to note contact inches away from the cat's head.

As whiskers are so prominent, it is easy to overlook the role of the rest of the cat's coat hairs as information receivers for the animal. The long guard hairs standing proud of the cat's coat seem to be of particular use in this respect.

Most mammalian carnivores have a common arrangement of facial tufts of whiskers. Besides the mystacial whiskers on the muzzle and prominent superciliary tufts of whiskers over each eye, the genal tufts behind the cheeks although present are less in evidence. However, unusually among carnivores, the inter-ramal tuft in the cat family is completely missing under the chin.

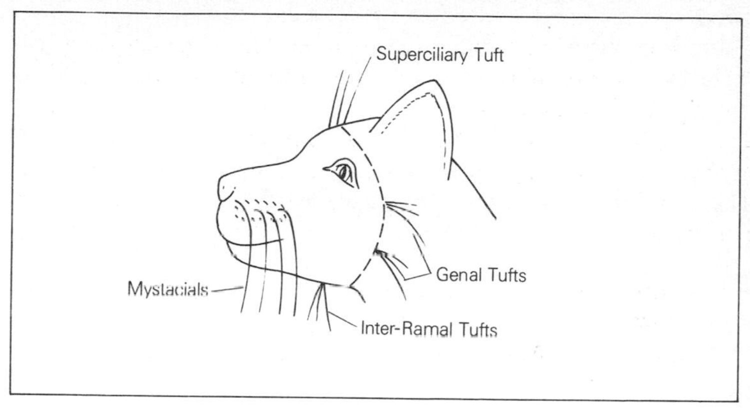

The general pattern of whisker tufts on an average carnivorous mammal.

It has been thought that as cats are relatively long-legged and do not feed by grubbing around, that there has been no need to retain this tuft under the chin. However as the stalking position of cats involves bringing the neck line parallel to, and within an inch or so, of the ground, it might be thought the tuft could have been useful. Its absence however is more likely connected to another aspect of cat's hunting – the kill. The requirement of the solitary killer is for powerful jaw pressure behind the canine teeth. By the aid of evolutionary pressure the cat family achieved this by considerably shortening the muzzle and jaw length. In consequence in the short space left an inter-ramal tuft would definitely get in the way for most normal activities, such as eating, washing and sitting (when it would be pressing against their chest).

Tufts of hair that probably do help the cat are often overlooked and these can be found on the underside of the forelegs above the paws. They are probably of most use in stalking to gauge footfall and possibly also in helping to judge a landing from a leap.

One perhaps unexpected use that has been suggested for the whiskers is as low frequency vibration detectors. The whiskers also keep the cat alert to air currents. Whiskers, like cats' ears, are a component of the cat's expressions. Far from being neutral objects the long mystacial whiskers on the cat's muzzle sweep into new positions, like a series of symmetrical conductor's batons to declare the mood and action of the cat. Whether the balling muzzle pulls them forward in a yawn or flat against the cheek when defensive, other cats can read the signs.

CHAPTER 3
Cat Communication

Cats talk to each other and cats talk to us. Of course the 'languages' are not the same but as you can get by in much of the world by talking pidgin english, so you can make do in the cat world by talking a sort of pidgin-cat.

It could be said that they don't talk – they make only a limited range of sounds that may have a stereotyped or specific meaning. However, that is the basis of man's languages, we just choose to call the sounds words. Some would say that the sounds are too general to have any meaning. But if you really listen to cats you would know that that is not true. Mildred Moelk in 1944 made a careful study of the feline audible vocabulary and distinguished sixteen utterances or phonetic patterns in the adult cat that could be linked to actual meanings. She found that unlike us the cat uses the inhaled as well as the exhaled breath for vocalising and vowel variation by changing throat-tension rather than jaw and mouth positions. She divided the sounds into three groups:

1. murmurs made with closed mouth
2. the fixed vowel sequence that is made from the open mouth gradually closing (aiou).
3. sounds from mouth held tensely open.

Certain calls seem quite unambiguous, such as the male and female calls that precede mating. Similarly, the deep distress howl that cats can utter at acute disturbance is unlike any other call in their repertoire.

The rumbling purr of contentment holds a fairly clear message, but it is not totally unambiguous. On anxious occasions a cat can give a more worried sounding purr, as sometimes encountered by vets when examining the animal.

Purring, like roaring, is something that separates our domestic cat, and the other small cats, from the big cats like the lion. Big cats do not have their hyoid arch totally made up of bone and so can give a vibrating roar, while our small cats' bony arch does not allow this. However, they make up for this by being able to purr both when inhaling and exhaling while the big cats can only purr when breathing out.

The spitting, snarls and shrieks encountered in cat fights also have a clear intention. So too do the low throat threatening growls of two cats facing each other at the edge of their respective ranges, glowering over the disputed grey area.

However, it is in their even-tempered range, rather than in rage that the domestic cat is more often heard by domestic man/woman. This more so than between cats themselves living a totally free existence. Certain clear phrases, such as the light trill increasing in pitch is used as a greeting by colony feral cats when two group members approach each other. This is accompanied by a slight arching of the back, a lifting of the tail and sometimes a bouncing lift of the front paws. This same communicating method is often seen by cat owners when they are met by their cats and the same sequence can also occur in response to offered food in the house.

The attention or food demanding meows (most insistent in the talkative siamese) and many of the low intensity grunts and mumbles, are more common in the place of hybrid conversation and in human habitation, than between totally free-living cats. Colony and farm cats tend to be quieter than domestic cats. House-cats note that our world is largely triggered by sound signals and so to make clear their meaning, and to prevent being overlooked, they increase the use of their voice. This occurs largely as a result of our satisfactory training *by* the cats to certain signals. Having found that calling out reinforces our attention when for instance they are sitting by the back door, they repeat this when they wish to be let out.

The emergence of a mutually understandable cat and human pidgin language is a consequence of joint conditioning, arising from the joint need to communicate. We have both conditioned ourselves and been conditioned by our cats usually to use the same phrase, such as 'Do you want something to eat', and when repeating the exact wording to repeat also the exact intonation which the cat recognises and acts upon. It is so easy to concentrate on body language expressed by the cat as part of this pidgin hybrid language and to overlook *our* repetitive movements that co-ordinate to a function recognised and required by the cat. Moving towards the 'fridge or picking up a saucer or tin opener while looking at the cat are examples of such human movements that the cat recognises, and then demonstrates its interest by its pattern of rubbing around your legs.

Sounds other than spoken sounds are important both to the free living feral cat and the domestic cat. The rustle of leaves and the squeak of a mouse are recognised by both. Both could clearly hear man-made sounds, but their meanings would not be the same to both. The rural feral cat may distrust the sound of all cars, while the urban feral cat uses the sound of car doors shutting and engine switched on to move from sitting under one car to sitting under another. The domestic cat can associate the sound of an individual car pulling up unseen, as bringing home members of its 'family' group. Similarly, individual unseen footsteps are differentiated by a waiting cat as 'family' or 'non-family'.

Combined languages

With our heavy dependence on audible language with its exact reinforcing sign counterpart (writing) there is an understandable tendency to think that posture, expression and intonation although useful, nonetheless are of a lower order of communication. However, the essence of the intention behind the words is often conveyed with great clarity by these very factors.

Our inherent species snobbery allows the assumption to be made that the limitations of language without audible sentences is greater than it is. We, as well as cats, make sentences of combined meaning, where the component parts are in more than one language mode. The cat has an array of audible sounds, many of which are applied on a specific type of occasion. These can be modified in intention by the involvement of audible expression and intonation. The rich language of body and facial posture can convey meaning by itself, but it is usually with another of the language modes forming a combined intention or sentence. These are further read in the context of position, such as whether clearly within territorial boundaries or near the limits, which in themselves have meaning from presence and marking. Confrontation position in itself conveys meaning. The sexual context itself is of significance. These, plus other factors, allow an expressive compilation that conveys a great deal of information.

What are the real limitations of such a method of communication? This should be placed within the limitations of the cat's intention which will stem from both functional requirements and capability. It is the level of capability that provides a stumbling block where species snobbery again is rampant. Many people believe that animals cannot have 'real' feelings or cannot 'understand' pain. However, it should be recognised that in a functioning animal appreciation of both emotion and pain are inbuilt from ancient evolutionary times. Emotional feeling has its believed seat in an early well-developed part of the mammalian brain and clearly from the survival viewpoint necessarily so. Fear, desire, rage and the others are emotional feelings of great assistance to a mobile hunting life form, and it is not a coincidence that it is well within the cat's repertoire to convey such messages expressively.

A language of movement, marks, sounds and scents that conveys feeling is particularly useful for the survival of a moderately social animal like the cat. However, its ability to convey identification of an object other than an immediately investigated one would seem minimal. In other words the grammar of the language contains a number of verbs. Pronouns are assumed in the language by intention, some adjectives of feeling are contained, but it would seem to be virtually without nouns except those of direction. This would seem to limit expression of concepts.

Expressive postures and following tails

Konrad Lorenz has commented that there 'are few animals that display their mood via facial expressions as distinctly as cats.' He noted that when a cat is becoming disturbed its growing alertness is seen in constriction of the pupils into a more slit-eyed shape, the ears lift and twist and so the cat's facial expression detectably changes. This may be accompanied by flickings of the tail. It is possible that the domestic cat uses nine clearly recognisable facial expressions and in addition sixteen distinct body and tail postures. To an extent these postures can be seen to form a cross-linking expressive series. It may be that the shortened faces deriving from similarly modified skulls offers a faint facial echo from man to cat, and cat to man. Certainly there are commonly identifiable facial and body postures when combined with correct use of inter-cat distances that I find make studying cat groups considerably easier.

From the normal expression (top left) Leyhausen showed moves to the right were increasing attack threat, and descending moves were increasing defence. Attackers swivel ears to show more of the back of the ear, while defenders flatten the ears sideways.

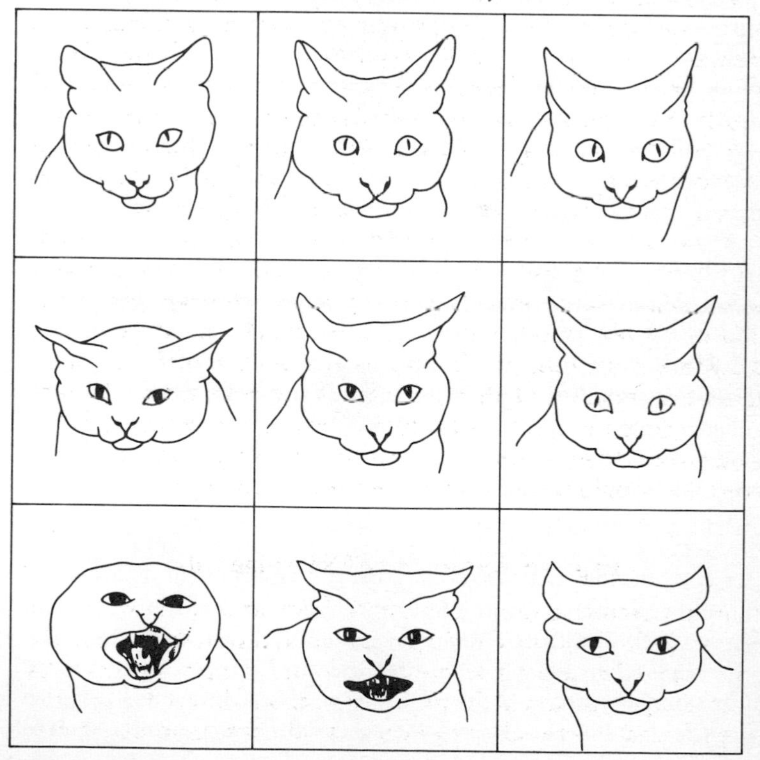

Following a cat or cats around London streets is fraught with difficulties. A suspect, if he becomes aware of being stealthily followed, is liable to bolt and not to go about his normal activities. Cats can quickly become aware that they are being followed and slipping from behind pillar box to telephone booth is not the answer. I have found that the most successful way is to inform the cat you are no threat and that really you are interested in something else, and that with their indulgence you will sit just peripherally to the group.

To do this you must be prepared to stand still at one place for a while at a distance to the group, looking around leisurely. Move in gradually, but not in the direction of the group, but at an angle to them, and looking away from them. Stop, still looking away, then slowly look in different directions, but still away from the cats. Then allow your head to move in their direction, and some of them will, although sitting still, be carefully watching you. Although in close movements with feral cats they will not seek out your face, at a distance of about twenty feet they will be scrutinising you to read your intentions. I will then usually blink, following immediately with a common expression. I will give a long-emphasised yawn in which during the latter part my top lip may curl to show more teeth and depending on the circumstances I may, holding my hands together, stretch my arms vertically downwards and hunch my shoulders while yawning. Initially I may even repeat this once or twice within 15–30 second intervals. With so much of their lives spent resting, cats yawn and stretch probably more frequently than less obligate carnivores, although much of this is done near solitary sleeping places, a fair amount of yawning and stretching goes on in unalarmed groupings. Yawning is therefore easily read as a reassuring sign, and some of the cats may even join in. It is then alright to glance around and gaze back at any cat looking at you.

However, after locking in on such a look it is essential to follow up with another reassurance signal. A slow and definite blink and hooding your eyes lower and lower, then slowly slightly open and close, open and close. This is a commonly used reassurance sign used between cats when they are lying in a hunched up sphinx position. Therefore if this is used while the cats are sitting up, looking for signals, this will be accepted. Then continue to look away with a head twist from the group while remaining stationary with the rest of your body, especially at a sharp sound, or anything that the cats' attention is alertly drawn to which should then be followed by small settling movements back on the spot, this movement being frequently used by cats in such circumstances. By now you will have become peripherally tolerated and the group will continue with its normal activities, but you must take care to be outside of the group's required distance, which is larger than individuals distances, unless you frequently turn up at the cats' meeting place at the same time as they do, when you will be recognised as less harmful. Alternatively, by accompanying a regular auxiliary feeder a relatively close approach will be tolerated. If you are not accompanying the feeder, sometimes, but not always, sitting down on the pavement to lose height in a hunched manner can be an advantage.

When following urban feral cats along streets it is essential to maintain no threat to their inter-animal distances or flight reaction will take over. The essential need for apparent disinterest in the cat however can cause the observer problems in an urban habitat. For example, sitting down on the pavement can arouse human interest which may disturb the cats. Similarly on one particular occasion I remember well I was engrossed in 'not looking' at a cat and I suppose to an outsider just seemed to be sauntering about slowly on the pavement in a studiedly random fashion, looking this way and that. When a house door that had been anxiously wavering suddenly fully opened and revealed a determined looking individual coming straight for me, I realised that for about ten minutes or so my movements had kept taking me very close to the side door of an extremely expensive foreign car parked by the roadside. My glances this way and that over my shoulder must have been misread by the car owner now coming towards me as a very different behaviour pattern due to the context in which he saw me, rather than the context I was presenting then to the cat.

In the course of my observations of free-living cats I have taken a large number of photographs and have found for instance that use of a flash light rarely worries them. However, a camera pointing at feral cats with you looking through it can be disturbing, but this can be prevented by the same expedient of looking around between shots. The need to look away to prevent the cat's responses focusing on the observer was particularly underlined for me when I was involved in filming for television a feral colony in London's West End (Appendix). A group of the cats were sitting around, some washing, waiting for the arrival of the lady who usually fed the group. I was automatically carrying out my 'non-commital' movements when I noticed the cats all stopped whatever they were doing and just stared. My eyes followed their gaze and I found that we were looking at the cameraman and his assistant, sound man, director, his assistant, who were all looking intently at the cats. Beyond this immediate collection were another eight or nine watching people, connected in some way with the evening's work, and beyond them huddles of more people looking to see what was being filmed. It was quite unnerving, a large number of eyes all staring, and no-one moving. It is small wonder the cats stopped as well. I quickly tried to reduce the number peering to a minimum, and the cats fortunately recovered their composure.

Calming reactions

On meeting strange domestic cats or when studying domestic cats moving freely outside, the ploys used above can also be used but the contexts are often different. On many occasions you are not such a neutral observer, especially if you live in the area, for your relationship to your cat and your resultant responses to other cats are noted and acted upon by the other cats. It is hard to remain neutral if pieces are being torn off your cat by an aggressive tom, and your intervention colours subsequent meetings. Similarly, if you are friendly with certain people nearby, you will probably find you are striking up a friendly relationship with their cat. Fortunately as most people do not wish to be dispassionate observers these events are not normally worrying.

Closer relationships and in many instances a removal of flight distances allows a greater use of 'pidgin-cat'. A cat will often settle down more quickly if you initiate 'foot-padding' movements with your hands by the cat. When a cat gives its rising 'Brrr' greeting call, lifting its tail, your semi-automatic response of stroking its back towards the tail is the best reinforcing friendly move you could make, even if you thought it out beforehand, for this makes the cat lift its back and tail even more, accentuating the greeting movement.

CHAPTER 4
Hot Cats

Have you ever seen a cat with sweat running down its fur? A hot cat, like a hot dog, would have difficulty in breaking out in a sweat, due to an overall reduction of sweat glands. The dog seems to disregard this and will run round as frantically in hot weather as in cold, but then collapse panting with wide gaping mouth and lolling tongue in a desperate attempt to cool down. This fits their hunting methods. Wild dogs will run their prey down while cats stalk.

One of the basic functions of sweat production is to cool an animal and in order to make the thin film of moisture on the skin evaporate, heat is taken from the skin, thereby creating a cooling action.

The cat, having a higher regard for its state of well being than has the dog, does not lightly undertake running or any unnecessary exertion in hot weather. Further more, great care is taken by the maintenance of a relaxed position not to overheat and thus keeping a steady, relatively low, metabolic rate. Body temperature control is then largely augmented by the cat's choice of temperature of its site. On a warm day it will carefully adjust its well-being by moving to sit in the sun to warm up and to sit in the shade to cool off. The general greater care the cat takes of itself, reaps its reward in having a proportionately higher life expectancy to that of the dog.

The loss of sweat glands in the cat does make such careful attention to environmental temperature essential. Anyone who has witnessed the great distress of an overheated cat that has, by confinement in a hot place, not been able to adjust its body heat by changing its surrounding temperature will readily understand this. The short fast wild gaping maniacal pant and staring eyes are not easily forgotten.

Cats do have a few sweat glands but these are localised to areas such as the chin and lips, around the nipples and between the pads of the feet. They are also mixed up with sebaceous glands at the anal glands.

One factor both ancestrally and currently that should not be overlooked in the man:cat relationship is that the transference of body heat and available heat is certainly something a cat values. The cat is happy to sit on its human group 'conspecifics', as against other cat group members when living wild. Farm cats living wild in Devon, when resting or sleeping spent about half of such time in contact with another cat (Chapter 5). When sleeping with group people or group cats, more than just warmth is involved. Company contact gives group reassurance and personal distance has been overcome. Non-group, non-related members are unlikely to make such contact. However if the rest of the human group accepts a stranger i.e. he comes into the house and sits with the hosts, then the house-cat may quietly sit on him.

The shape of a cat when resting has been found to be a very good indicator of the difference between the cat's own body heat and the surrounding air temperature. The tighter the ball into which a cat curls goes with cooling air temperature while a languid stretched out cat demonstrates warm surroundings. This can be seen whenever a cat sleeps by a fire or sprawls out, tummy exposed, to the sun. The ¾ and ½ circle positions can be seen between the extremes.

Despite this the cat is fairly insensitive over much of its surface to contact of hot or cold. It can tolerate temperatures that are painful to people and consequently will sit and walk on a 'hot tin roof'! The cat's nose and upper lip are however very sensitive to small changes in temperature and could be seen as the cat's external thermometer.

Weather prediction

The belief that cats can foretell the weather has been handed down to us in folklore sayings. However, this stems from an earlier belief prevalent at the time when cats could be presumed to be witches' familiars, that cats had power to affect the weather. The fineness of the dividing line and the ease of transition between the two is reflected in the Irish saying that putting a cat under a pot brings on bad weather and in consequence was sometimes done 'tongue in cheek' (at least by Victorian times) to induce a guest to stay. Most weather sayings concerning cats could be read in either light depending on the prevailing sentiment.

As the origins of witches lie in the weather/fertility rituals of early man, it is perhaps natural that cats should have been considered to have weather control by association. However, many weather ditties doubtless arose by observation only seeking prediction. The cat, being a household animal, is easy to watch. In this connection probably more weather sayings exist about cats, horses and cows than about any other animals.

> 'Careful observers may foretell the hour
> (By sure prognostics) when to dread a shower,
> While rain depends, the pensive cat gives o'er
> Her frolics, and pursues her tail no more'.
>
> JONATHAN SWIFT

Richard Inwards collected weatherlore throughout the second half of the nineteenth century. He recorded the following lines:

> 'When cats wipe their jaws with their feet, it is a sign of rain, and especially when they put their paws over their ears in wiping'.

Although many weather maxims seem somewhat esoteric this one accorded with my own observation and it is also clear in this instance that there could be a biological basis for it. I discount the washing of the jaws, for in my experience this occurs after virtually every meal and I can therefore see no tie up to weather changes. However, every time I have ever known a cat not just to go over its ears in washing, but also firmly rub into the little depression just in front of each ear, it has resulted in rain by the following day. I am not suggesting witchcraft, but merely that the inner ear and eardrum are sensitively attuned to detect slight air pressure changes, for this is how the vibrational nature of sound makes itself felt. To a lesser extent humidity changes will affect the membrane-free surface.

We are certainly aware of similar air pressure changes when we take off in an aircraft, or even when taken up in an enclosed very tall building, by 'popping' in the ears and a distortion of sound quality, and it would be likely to be something of this nature that cats could detect on the unseen approach of rain. We would only see the paw rubbing of the ear when the change is irritating (and we happen to be looking). Certainly as free living cats abhor abandoning shelter for food in heavy rain, as a saturated coat aids neither comfort nor survival, some earlier intimation could allow expedient action and would seem to offer certain survival advantage.

All of this I had accepted as at least being plausible; I therefore found it slightly disconcerting to find a contradictory rhyme from the northern counties:

> 'If the cat washes her face o'er the ear,
> Tis a sign the weather'll be fine and clear'.

I was therefore slightly reassured to read in the poem on rain prediction by Dr Erasmus Darwin (grandfather of Charles) the lines:

> 'Puss on the hearth, with velvet paws
> Sits wiping o'er her whiskered jaws;
> Twill surely rain – I see with sorrow
> Our jaunt must be put off tomorrow'.

Slightly perplexed I conducted a survey via 'Animal Magic' (BBC 1) without giving any bias to either of the conflicting tales. Those taking part were asked to notice if and when their cats washed over their ears and to note how long afterwards it next rained. It was asked in this form on the assumption that if there was a large scatter of results, or that it rained at a much later date, then there was probably little correlation to any rain prediction. As observation was not continuous, most missed the ear rubbing (it only lasts for less than a minute) or realised it was raining and that they had forgotten to note when it started. The few that noted both at least showed it to be an only occasional piece of behaviour (it must be one of the least washed parts of a cat's body) but the results did seem that the cats may be acting as barometer.

The results, such as they were, were tightly grouped, and it seems possible that cats predict rain a day before it happens, but that most ear washing occurred within four hours of rain, especially in the final hour. Now whether the cats interpret the stimulus to the ear washing response as forthcoming rain is open to debate, but I see no reason why they should not, as such interpretation could be of selective advantage. A cold and wet kitten is a dead kitten. There was no north/south difference in the results, so the southern line seems to have won out which does fit with my own casual observations of cats' ear washing behaviour over a couple of decades, when I have always found it foretell rain within a twenty-four hour period. However, I do have to admit to having lived for most of my life in the south of England.

Although this is far from being the most convincing survey ever held, it does add the tiniest bit of weight (perhaps in this instance the apothocaries weight of half a scruple might be most apt) to an area of the cat's life and times that intrigues me, for the overlap of weather, witches, survival pressure and behaviour makes a potent cauldron's brew.

CHAPTER 5

The Social Cat

Paul Leyhausen, a few years ago, drew attention to the possibility that many hitherto believed solitary mammals, particularly cats, could have some degree of social structure, and believed that this was usually overlooked from an absence of field work on these animals. He strongly suggested that even largely solitary animals could nonetheless fit into a social pattern.

It has been suggested that mammals do not so much live in an area as on a series of pathways connecting key interest places including particularly for cats sunny sleeping places. For many mammals a clear distinction has been drawn between the animal's home range and its territory.

The home range can be thought of as the area that the animal lives in and the territory can be thought of as that area that the animal normally defends. Consequently the territory for most mammalian species (that have a territory) is smaller than the home range. When looked at like this it seems quite simple, but that is misleading, for of necessity both home ranges and territories that are known follow observations of one kind or another that often have some limitations, so that such ranges are recorded ranges rather than the real ranges. Range size varies with habitat, and particularly with availability of food and in this sense there can be no 'true' size, but only certain limits. Leyhausen believed that free-ranging domestic cats behave so inconsistently that it was not possible to make the clear distinction of home range and territory. He noted an over obsession among animal behaviourists to study aggression that put the concept of individual territory on a firm footing. However, space can often be better used than in an

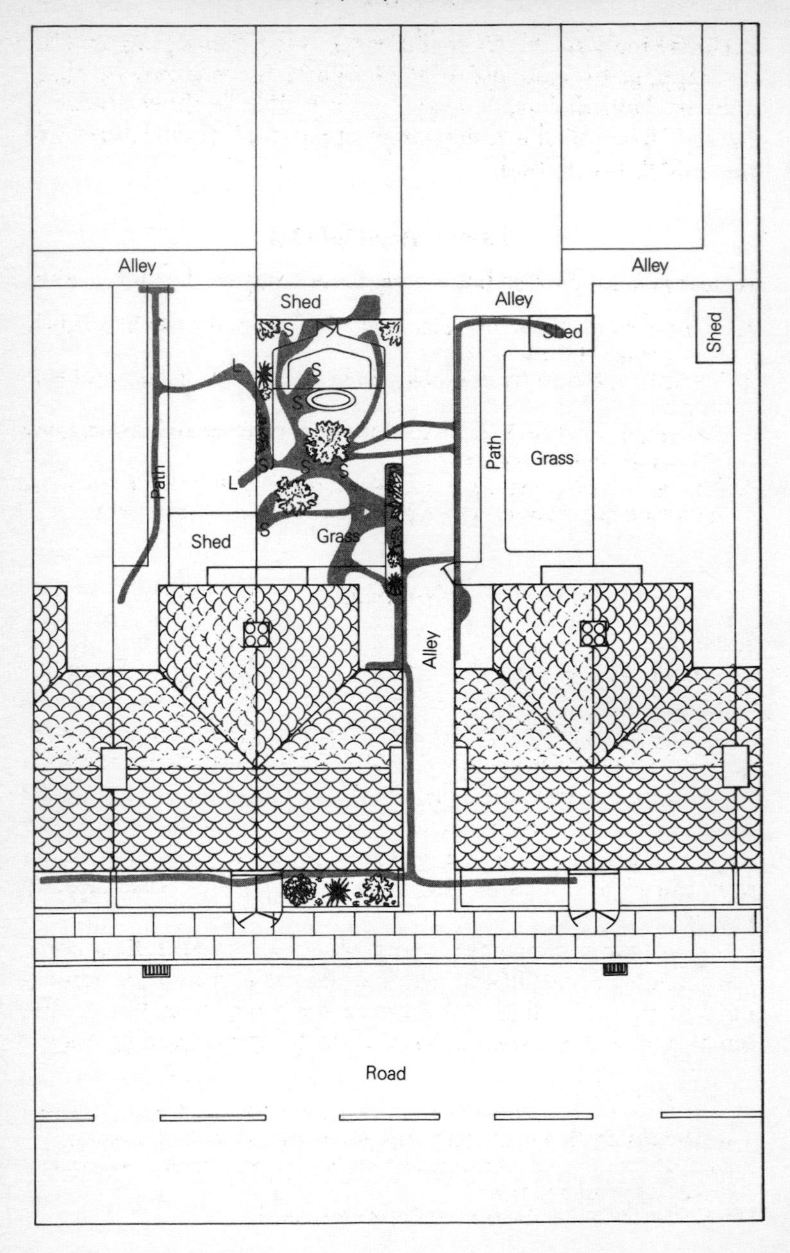
Alley
Alley
Alley
Shed
Shed
Shed
Shed
Path
Path
Grass
Grass
Alley
Road

exclusive individual 'ownership' way, so social groupings and the overlapping of land use often modifies the concept in many animals. Nonetheless, domestic cats certainly do show a range of sizes of an identifiable home range, apparently relating directly to the availability of food.

Home ranges for cats

A clear general pattern has emerged over only the last few years:

1. A male cat uses a home range around 10 times the size of a female home range in the same area.
2. The greater the degree of supplementary feeding the smaller the home ranges.
3. The female cats' home ranges are clumped together and considerably overlap around a core area.
4. The males' range include a number of queens, and a male will mate with both queens of his prime core area and of other core areas.

Case studies

Farm Cats

David Macdonald and Peter Apps in their study in 1978 radio-tracked a small group of farm cats in Devon and their findings were consistent with the above general pattern. The cats were largely self-supporting and so their ranges can be taken as a measure of the area of English landscape required to provide the food needed without much supplementary feeding. Nonetheless the scavenger part of the cat appeared, for the tom not only caught prey but raided dustbins and ate carrion such as road casualty rabbit.

The Devon farm females were living close together, their home ranges being about fifteen acres, while the tom ranged, particularly to the outer limits of an area some ten times larger. The females stayed grouped around farm buildings gaining benefit

---

The path system of a neutered female house cat in the Barking study area. Wider shading denotes greater use. S = sunning spot, L = latrine (latrines are semi-peripheral).
(The cat lived in the house with plants illustrated.)

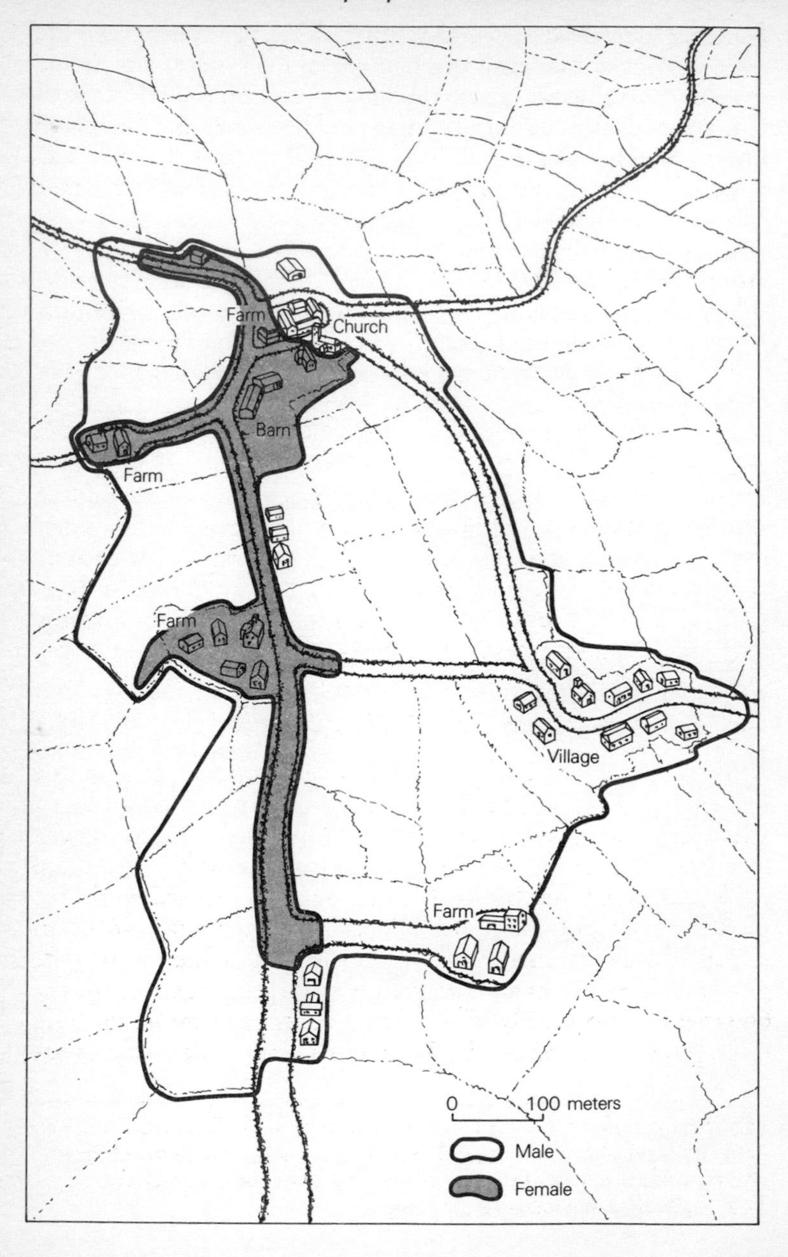
Farm
Church
Barn
Farm
Farm
Village
Farm
0
100 meters
Male
Female

from shelter and associated small rodents. (A recent study of American farm cats in Wisconsin by John Laundré, also found that the cats using the farm buildings for hunting and resting were females, while males ranged further afield). As well as mating with his group of females during the course of the study the male also visited and successfully mated with a cat from another farm at the other end of his range.

It has become accepted that the English rural cat density is around one cat per 20–25 acres. Overall this would seem to be the same density of cats in the Devon farm study, (although only about one cat per fifty acres if just the adult group members are considered). These values are within the range of one cat per twenty acres down to one cat per 250 acres that have been found for a number of far-flung rural studies (America, New Zealand and various isolated islands). (If England were entirely rural then it is possible that its dry landmass could support about 1¼ million cats living off their wits without additional feeding, and similarly the UK overall some 2¼ million. These figures are not entirely spurious, for they indicate our current rural population of free-ranging farm cats and free-living feral cats might be something like three-quarters of a million or even approaching a million for England, and double these estimates for the UK).

The dockland cat

A very fine and detailed study has been carried out on the movements and home ranges in daylight and dusk of a group of 200 adult feral cats by Jane Dards in Portsmouth Dockyard. The location had certain advantages for such a study, the main advantage being that the dockyard is an almost closed world for the cats, being some 210 acres bounded by the huge old naval dockyard walls on three sides, and by the sea on the fourth. The only unaided ways in and out are the few manned gates. One cat however became locally notorious for using one gate for its daily return visit out to the nearby fish and chip shop. The coat colours

---

Home ranges of the tom and one of the queens of the Devon farm cat study by David MacDonald and Peter Apps. The core area of activity and base of the cats was the farmyard area at the top of the picture. Tom regularly scent-marked his range.

of the cats testify to the isolated stability of the population. Considerable auxiliary feeding of the cats is undertaken in this case by a number of dock workers. The cats also enveigle fish from casual line fishermen by sitting in sorrowful groups around their potential provider. (Mice and rats are controlled by poisoning.)

Absolute definition of home range is difficult in dockland with interdigitating fingers of water and land, and multilevel buildings and culverts, with the additional problem of dry docks that become a water area during use. Nonetheless, by computer interpretation of over 4½ thousand sitings Jane Dards established home ranges in the land area for ninety-five cats. There was variation between individuals, but again generally the males had a much larger range than the females, about ten times the area. This could be for instance twenty acres compared to two acres. A group of cats would utilise the same core area. The small female ranges focused on core areas, while the toms' larger ranges overlapped several female core areas. Before maturing young toms would live in the female core area. In terms of density Jane Dards found one cat per acre in the dockyard.

The population in the dockyard was maintaining an apparent stability in the years studied. While some 400 kittens were being born each year, only fifty reached adulthood. Also some fifty adults were dying or being killed each year, largely as a result of road accidents. In consequence, if the level of cats is acceptable to the dockyard personnel, attempts at control would not seem necessary, and a balanced state would exist.

Urban cats

Urban feral cats' home ranges also fit into the general pattern. The size of the ranges do depend however on where they are. Hospitals with large grounds allow a larger potential size, but in our cities

---

Portsmouth naval dockyard (1.6 × 1.3 km) is enclosed by high dockyard walls and the sea. Jane Dards mapped the position of cat-group ranges, plus feeding sites where cats were fed by dockyard workers at least once a weekday, and skips and bins scavenged by cats.

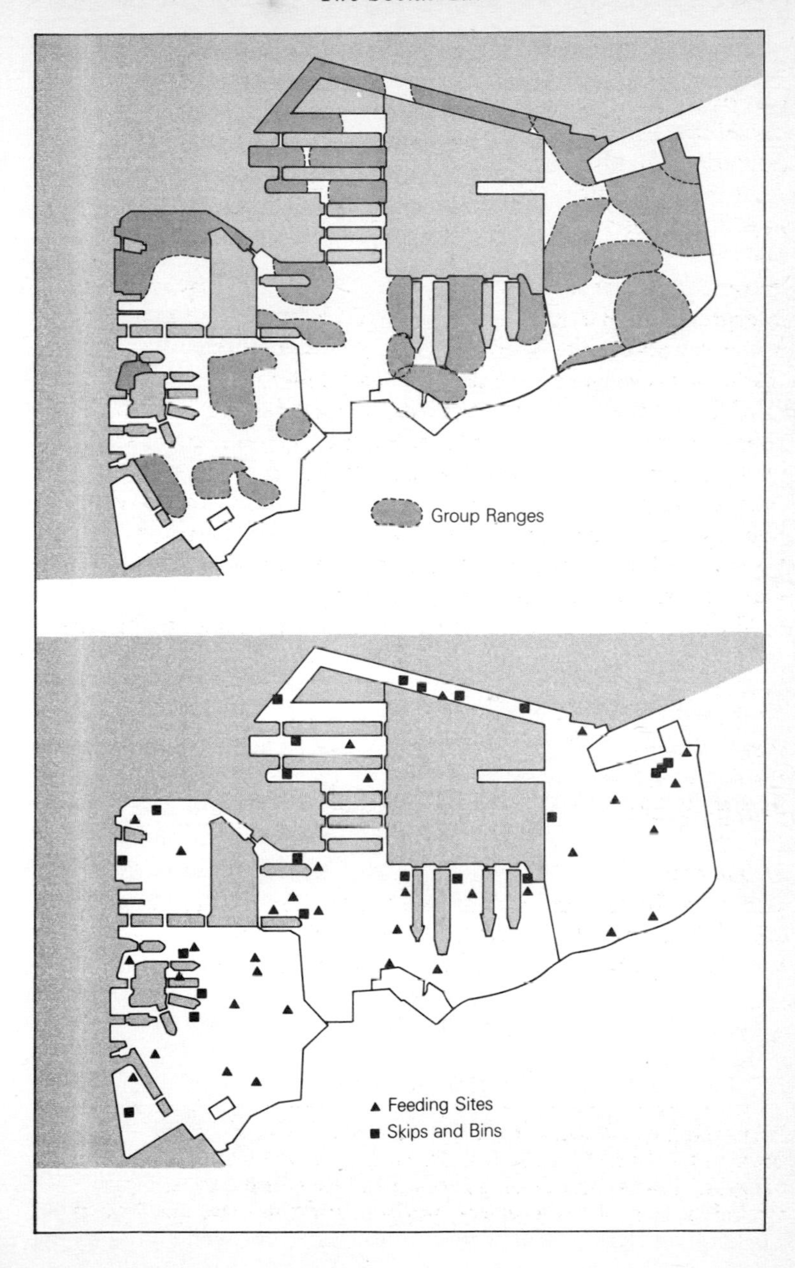
Group Ranges
Feeding Sites
Skips and Bins

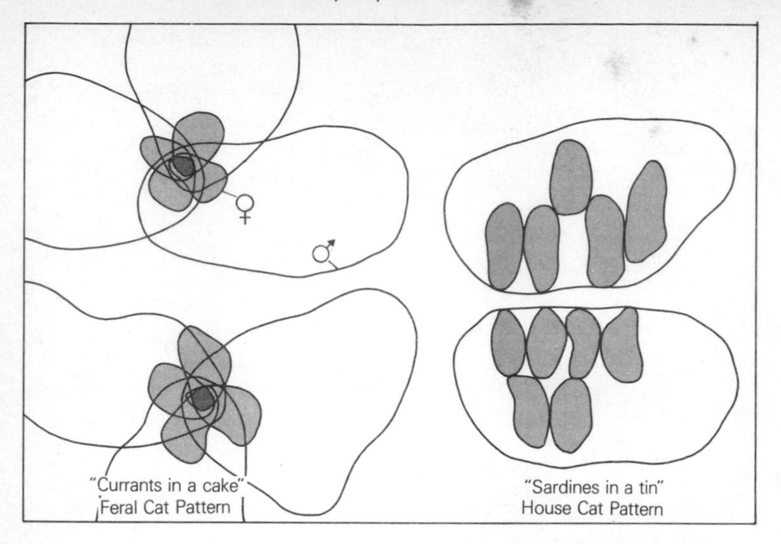

Feral cat groups have female overlapping ranges enclosed by male ranges. The house cat pattern has non-overlapping female ranges enclosed by male ranges, for these cats have formed groups with people.

urban cores space, as in London, is at a premium for both man and cat. In the 'high density arc' of central west London, where railings provide shelter, scavengable food is accessible, and generous auxiliary feeding provided by a battalion of feeding ladies, then the feral population has a density around one cat per 0.2 acre, (although outside of the arc it tends to be lower but other localised high density areas exist where conditions suit). The core areas are found in railinged squares, dimly lit back street car parks and similar places.

The female cats of Fitzroy Square (Appendix) in the 'high density arc' have home ranges of about half an acre, with the toms again around ten times the size. Both sexes clearly utilise the core area.

The general pattern for all the feral groups I have visited has held, and males used a greater area than females. Although visiting other 'colonies' the toms of Fitzroy Square nonetheless maintain their base within the one group. The high density housing of central urban London with effectively no individual private garden area attached to houses is not well suited to the

Above: Forest (Scottish) wild cat, the most likely candidate for ancestor of the modern domestic cat (see p. 12)

Right: Domestic living has enabled a small percentage of cats to well exceed the average life span. Alert and lively at 20! (see p. 164)

Above left: Many islands have cats, but survival is not easy on the arid equatorial Galapagos Islands. Kitten seeking shade (see p. 150)

Below left: The small lava lizards are a major prey for the Galapagos Islands' cats (see p. 128)

3/4.

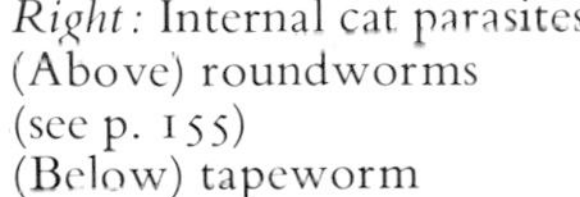

Right: Internal cat parasites
(Above) roundworms
(see p. 155)
(Below) tapeworm

Below: Galapagos Hawk with its kill, a feral Galapagos' cat (see p. 152)

Above: Both house cats and feral cats love the warmth and security of cardboard boxes. Cats living wild in East London (see p. 84)

Right: Recently strayed cats remain approachable

Facing page: The cat has spread worldwide, yet density depends on food. India has so many species of competing scavengers that there are relatively few cats. Few are tolerated as pets, but these at Jaipur are appreciated

Above: Author scanning a factory feral cat for ringworm

Right: Fitzroy Square cat feeder about to release the colony cats after neutering. (The blanket cover decreases their anxiety during handling) (see p. 187)

Above left: Newly formed groups do not seem to have the stability of long-established colonies. These young cats died of infection before becoming adult

Below left: When unrelated cats are penned they carefully maintain a personalized space (see p. 74)

Above: Miss Mary Wyatt feeding some of the Fitzroy Square cats in the evening (see p. 195)

Left: Co-operation between biologist and cat auxiliary feeding ladies has helped a better understanding of the urban cat

ranges of domestic house cats, for the householders do not have an identity for the cat with squares. The squares for the same reason ideally suit the feral cat groups, but consequently domestic cats are fitting around the feral cat pattern – the reverse of Suburbia.

The suburban cat

In contrast to the other groupings above, the domestic house cat living in a house with a garden, at first sight does not fit into the general pattern. Yet unlike dogs, hamsters, gerbils or any other domestic animal, the house-cat is entirely free ranging and once outside of the house it would seem strange if it suddenly invented a different home range pattern to the completely free-living groups of cats.

As shown in the East London suburban study I carried out at Barking from 1976–1978, and this seems to be typical for household cats, it is the female cats which have a home range of the house and garden plus any further space the cat density will allow. When a female cat from the London suburbia study area was removed and placed in a village cottage its home range grew, but only as a reflection of larger plot size. A series of parallel female home ranges are seen with housing. A typical female home range for the dense suburban area was 0.05 acre. Males however still use a home range approximately ten times a female range. Intact toms appear to have a slightly larger home range than neutered toms.

The cat density in the East London study was one cat per 0.02 acre, the highest density of the studies, and one thousand times higher than the feral rural population.

The females used their home ranges in much the same way as did the cats of the other studies, for even in these small gardens certain spots were consistently identified as greater use areas, such as sunbathing nests. These areas were linked by routes that were walked more than other areas, but it is doubtful if in these small home ranges any area could be as clearly defined as 'none-use', unlike in larger home range cats.

Although female cats occasionally met other female cats, the normal situation was one of strict territorial observance. In very small suburban garden areas while the region that was territorially defended approximated to that of the home range, it was always nonetheless smaller.

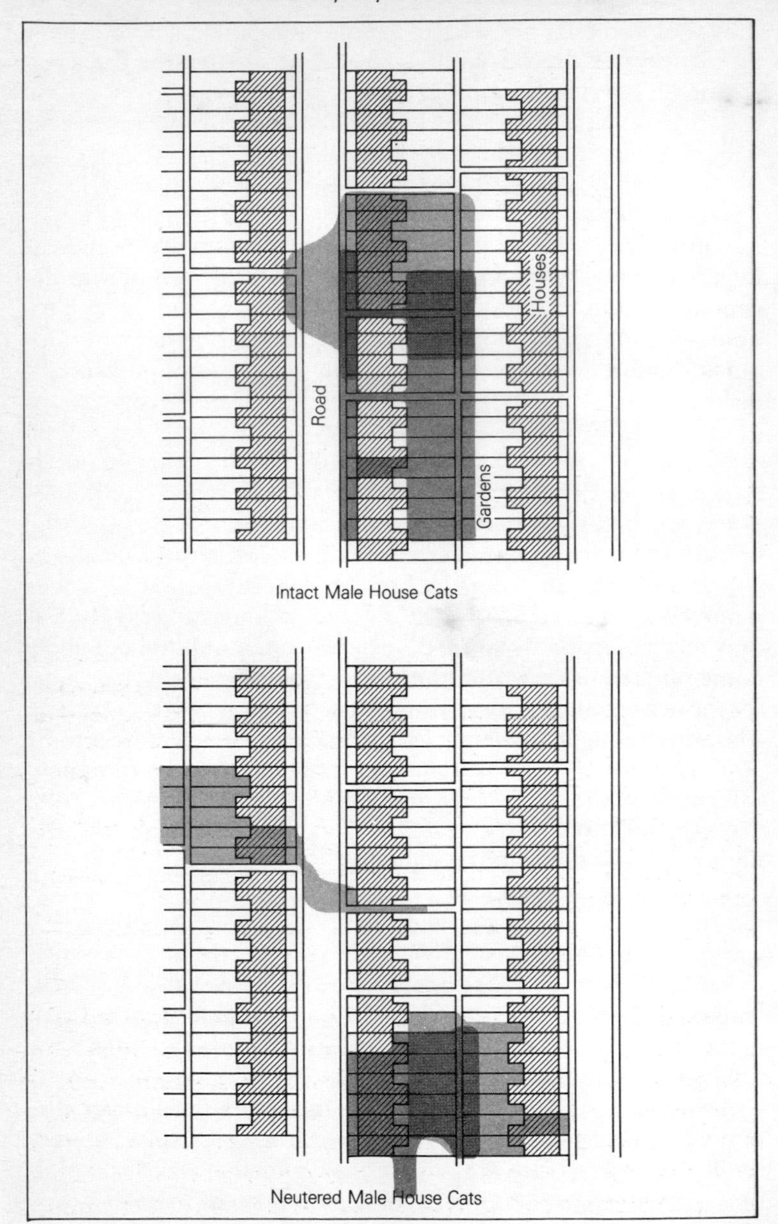
Houses
Road
Gardens
Intact Male House Cats
Neutered Male House Cats

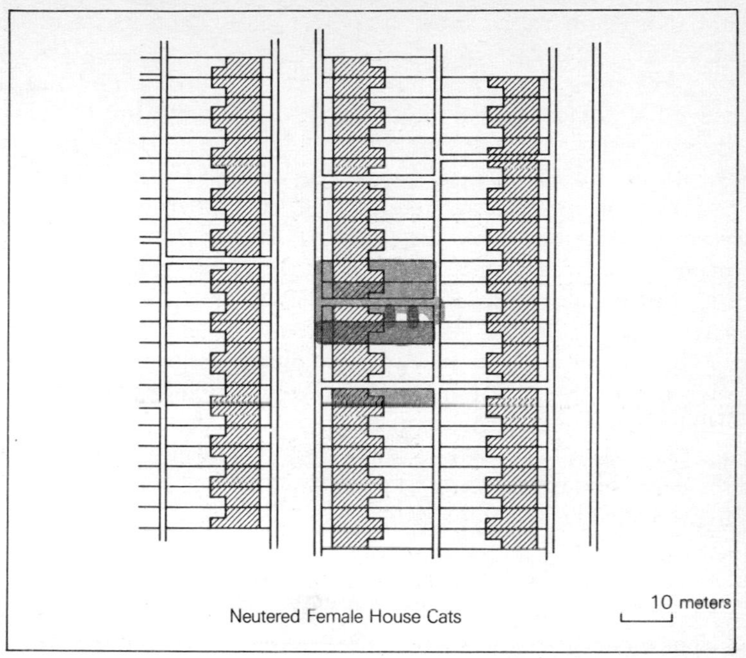

House cats are free-ranging, not being contained, and like their feral counterparts form their own home ranges. The maps show part of the Barking study area in East London's suburbia, an area of terraced housing, small gardens and alleyways behind the gardens. Each of the three maps shows the same group of houses and gardens, so each map can be superimposed on the others directly. Only a few cats of each type are shown to avoid confusion. Queens from different households overlap minimally. Toms, neutered and intact, overlap queens and each other in their larger ranges.

Female feral cats have a mutual core area that is often linked to a point of high food availability. If the humans in the house-cat's life are seen to be considered by the cat as surrogate cats, its 'conspecifics', then much becomes clearer. The house-cat's behaviour towards its human 'conspecifics' seems to bear out the view that the cat sees its people as group members. The house-group of humans and female cats have overlapping home ranges and both noticeably mark their territorial limits. Man-made structures of car-carrying roads, houses, sheds, hedges and fences play key roles in delineating the used area. Area confidence is gained by the cat not by these physical structures, but how her group uses the space. Cutting the grass, clipping the hedge, tidying the border or similar activities are carried out in an identifiable area that the house-cat recognises as the cues for group range. Her core area is in the house where feeding also occurs. Consequently neighbouring female cats, each with their own group and core areas are not of her group and home-range overlap is minimal and territories respected as between different feral female groups.

The pattern is also a recognisable one to the tom. He has his own family human/cat group and the collection of parallel suburban female home ranges must seem like females of other groups that he includes in his home range. Just like the farm tom visiting the peripheral neighbouring farm female cat named 'Broken-ear' in the radio tagged Devon study.

Male buffers

The home range pattern produced by feral cats in the above case histories all produced discrete female cores buffered from others by the males' ranges.

The far greater size home range of the male cats over female cats cannot be due only to differences in food requirements. The male may have a higher energetic requirement, but this is unlikely to be sufficient in itself for the difference.

This buffering probably makes for social stability for a genetically related group, while at the same time allowing, via the toms, an essential gene flow through peripheral matings. This land pattern plus the toms total avoidance of paternal role probably also reduces the spread of kitten bronchial diseases. However, the suburban or garden pattern of home ranges loses any buffering as the females pack into the toms range like sardines in a tin. Despite this, the tom:queen home range ratio still holds as the females do not overlap.

Domestic cats could not easily identify their human home range in London's West End high density gardenless housing and were set around a feral pattern. In contrast in suburbia the domestic pattern is supreme, the feral and stray cats have to fit colonies into gaps in the housing pattern caused by factories, railway yards, hospitals and similar places. Any strays and ferals entering the tightly packed suburban housing pattern cannot form groupings and adapt or move on after a few months. For example over a five year period a half acre zone of the Barking, East London suburban study area was seen to be transiently used separately by four strays for periods up to some months, and one of these was taken in and homed within that area. However they all had to exist within the social conventions of the suburban domestic cat population.

Territories

Your cat's territory is not a constant area, but fluctuates in size from time to time. Taking an adult domestic cat's movements over a year it is apparent that there is both a mean home range and a mean territory size. The cat's willingness to defend the area depends upon its confidence with regard to the cat it faces following the experience of their previous meetings. A flexible boundary exists against different intruding cats at any one time. I have known a neutered female to offer no defence against an aggressive tom in any section of her home range, but instead hide away, whilst in the same day standing firm near her normal territorial limits to the adjacent female neuter. Previously (and subsequently) the same home range occupant had held similar limits for both cats.

Rather than say the application of the concept is too variable it would be better to have a measure of the variation. It is possible either to take current territorial measures as a 'rule of thumb', or better to place less emphasis on a line that can be drawn, but to take an image from nuclear physics and say that the line has a probability of existence at any one point and to conceive of it as a blurred edge. House cats have a particular significance for such studies as virtually all of their food is provided, so home range and territory sizes are due to social factors.

Why are home ranges different sizes?

When looking at figures for home ranges of different species of animals, it is easy to think that they are some sort of absolute values for those particular animals. Far from it, home ranges vary due to many factors such as time of year, weather, food availability, animal density and so on. Nonetheless for most animal species variations are not too large, and indeed different methods of measuring ranges employed by researchers are frequently believed to have as much inherent variation between them! Most variations of home range for any species of mammal usually comes down to energy availability.

The body weight of any mammal, (including your cat) is strongly tied to that animal's rate of energy use which naturally is related to the amount of food the animal needs. It is normally the availability of this food that strongly dictates minimum requirements for home range size. It has been found that it is possible to relate the size of mammals home range directly to body weight.

Animal species that have to hunt around for their food (and this includes animals that for example have to look around for nuts, as well as the obvious hunters such as the big cats) have larger home ranges than browsing, cropping feeders, where food is readily available. When the figures for cats' home ranges are considered in this context, their body weights, feeding methods, behaviour, density, and home range size can be seen to form a coherent pattern.

The home range found for some rural feral cats can be seen to be the same area that would be expected for a free-living mammalian hunter of their size. In other words they are performing as wild cats. The home ranges encountered by Jane Dards in Portsmouth dockyard span a range from similar sized male ranges to considerably smaller female ranges. The generalised graphical treatment does not allow for large sex differences and ratios, nonetheless cats whilst largely obligate carnivores can be seen to be covering the 'croppers' range! This shows that they must be getting quite a lot of additional feeding or very ready access to scavengable food (in the dockyard). When the domestic cat suburban home ranges (Barking study area) are looked at it is immediately clear that their home ranges are well below the area of not only hunters, but grazing animals as well, for their size. This can only be considered possible due to food being so readily available. The household 'auxiliary' feeding has enabled your domestic cat to become a 'super-cropper'.

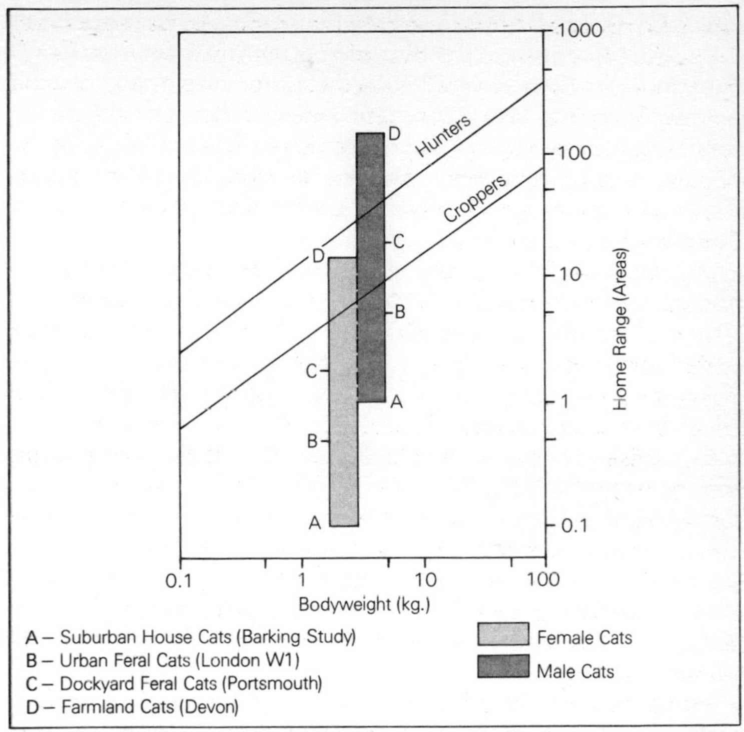

Small bodyweight animals do not need such a big home range as large ones. A hunter (top line) that has to search for its food needs a larger area than a cropping animal (lower line) of the same bodyweight that has its food always available. Brian McNab showed the direct link between bodyweight and home range for hunters and croppers. Now it is possible to fit the recent domestic cat studies to this pattern and gain some insight to their way of life. The farmland cats require the home range a hunter's life demands (D). The greater amounts of auxiliary feeding, through the urban ferals, leaves the almost totally saucer-fed house cat with minimal home range requirements – 'super-croppers'! (A).

However as they are carnivores they do not have to feed continually in the manner of a sheep or rabbit, for the carnivores' food is nutritionally richer. Therefore although, for example, a lion requires a very large area to hold sufficient prey items for both them and him to survive in, when a lion pride makes a kill the food that can be gulped down from the carcase in a couple of hours will keep the mechanism of the lion ticking over for a couple of days. Consequently lions have a lot of spare time which they sensibly invest in sleeping. This is the same course of action taken by both domestic and feral cats. Your domestic cat, not needing to spend the same involvement in hunting or scavenging that the rural feral cat does, has the opportunity of spending even more time resting and sleeping! It may be this attitude to life that gives them their higher size-relative life expectancy over the more omnivorous and active dog.

Born free – but comfortable

Most animals are either domestic or wild, or perhaps it would be better to say captive or free-living. The domestic cat is almost unique in maintaining a free-ranging life while being apparently domestic. They are sufficiently free-ranging as to be able to choose their own home range in the manner of a totally free-living animal. Farm stock, although free to move, are contained by field size, pet rodents restricted by cages, and although the free-ranging 'latch-key' dog does exist, its numbers hardly compare to the numbers of normal domestic cats except in places like India.

For most free-living animals food availability has largely dictated home range size with regard to metabolic need of the size of the animal. The domestic cat with its super-abundance of food has little home range requirements, allowing the cat hyper-density of suburbia possible. (Just as well or far fewer people would be able to live with a cat!)

The cat's uniquely ambiguous free-ranging yet domestic status, which ensures people trying to feed it even when it lives a totally free life, has allowed it to live over a greater range of home range sizes and over a greater range of densities than any other predatory mammal. From the four studies outlined in this chapter, the more auxiliary feeding the cats have received, the less home range they have needed, and consequently their densities have been able to increase. This relationship seems so direct that it should be possible to obtain a guide of home range size for any population of cats from a knowledge of the cat density (and vice versa); – and of course some idea of the amount of food they are getting that they have not caught.

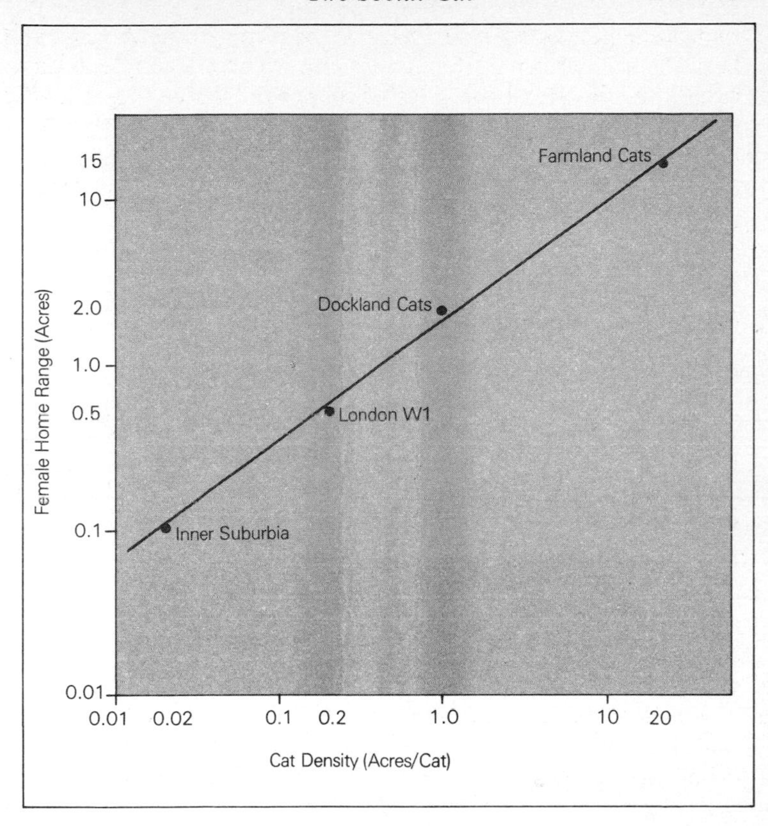

Of the many factors affecting home-range size it now seems that food availability is the major one. In animals, like tigers, the type of habitat affects the amount of prey that the habitat can carry, and large differences in home range are not apparent. The domestic cat, due to its huge span of home-range size, seems to be a powerful model for other carnivores, like the big cats. In the domestic cat the rural feral cat (top rt.) requires a large home range to support its prey, the dockland and urban feral cats (mid) receive greater amounts of auxiliary feeding from a vast army of feeders so do not need so large a range (New York urban feral cats in a study by Carol Haspel also fit this mid range). Saucer-fed house cats have minimal home-range needs. As home range size decreases, so cat density increases in a remarkably direct manner.

Cat distances

One of the indicators of the length of established ties in colony cats is the inter-cat distance. It seems that the long-term established feral colony (of close family ties) is more akin to the suburban human/cat household 'colonies' than in either short term colonies or between cats from different houses of suburbia. Adult cats from either of the latter two will normally sit no closer than a certain tolerable distance. There are however qualifications. First, kittens will often play together and rub heads even in a new colony, especially if feeding time is near, but this is more likely to be between kittens of one family. (Similarly kittens will involve in mock fights and 'seriously' engage in determined stalks and pounces on such savage prey as a paper bag!) In adults however distances are strictly maintained, subject to overriding events such as mating and fighting. However, in normal conditions a minimum distance is maintained which if broached may trigger a pre-fight or flight display pattern. If a new colony is assembling with a knowledge of approaching feeding time, then the local cat density may typically be at such a time twenty cats per 100 square yards. We practise this selection of equal spacing ourselves when we are away from our home colony, such as on a bus or train; we tend to select empty seats if possible a reasonable distance away from other taken seats. For us too our interpersonal distances are less on our home ground, but even then it is dependent on the closeness of bonding.

When non-related cats have been placed randomly in pens I find this distancing is carefully maintained, even at very high densities so that the cats may only be two feet away from each other all round. These distances are still maintained even when the cats have been living together for some months. Friction only usually occurs in these groups when distances break down over feeding.

Cats in colonies that have been established for a long time, although still tending to maintain distances when living free, even when anticipating feeding, will nonetheless enter into close contact and rubbing of heads in a similar manner to household cats. When however I have taken such a feral colony and placed it in a pen away from its site in complete contrast to the penned random cats, no attempt whatsoever has been made to maintain distance.

Instead the group have voluntarily placed themselves in close body contact with other group members. The group's identity is maintained in a strange setting with a group confidence. In such a long-established colony all of the cats are related in one way or another to each other, and the greater social support is of direct value to the gene group.

The house-cat's greater degree of social interaction to what are not only non-closely related group members, but are not even members of the same species may be in response to supernormal stimuli. Gull chicks for example respond more to large red blobs than to the natural size red mark on the feeding adult's beak, and the 'Playboy' empire was built by making out that the supernormal pneumatic image was normal (even if dressed up like rabbits!).

House-cats receive a superabundance of food and when interpersonal distances are broken a superabundance of warm body contact with much rubbing (tickling and stroking) instead of just the brief head rubbing with another cat. So perhaps we should expect that a hyper-response of social activity occurs to these supernormal stimuli.

Rank aggression!

Certain trends over the years, in the course of animal behaviour studies, have cast light on particular aspects of activity. However there is a great danger of producing a false overall picture merely because other aspects have not been studied. The particular studied trends could even be taken as giving more insight into us than the animals observed! Following the formation of the concept of territory for birds, extensive efforts have been made to establish this for most animals. Similarly extensive studies have been made of aggression and social ranking to be universally applied. In comparison, studies of affection and similar socially cohesive forces are sparse. A selfish gene does not of necessity mean a competing gene.

I have found aggression in cats, but more often I have seen affection between free-living cats. Behaviour studies usually depend upon obvious identifiable features that can be counted and perhaps aggressive acts are easier to identify. Positive cohesive features that are initially less identifiable have often been largely overlooked. These include for cats such positive actions as the group of cats sitting quietly near each other, even though maintaining a personal space. This usually occurs for most feral groups around the feeding area, but I have seen a group of twenty or so cats congregated during the evening under a car park ramp before individually moving off into the basement of a nearby hospital for food.

Dominance among cats?

Over the years many studies in animals have been aimed at establishing dominance-submission relationships in groups. These have usually been carried out under laboratory conditions 'to minimise the variables'. The assumption that putting animals under laboratory conditions only reduces the number of variables to which the animal is exposed is clearly suspect. The conditions apparently simplified may change the behaviour of the subject. The effect of caging animals usually reduces home range so much as to make the animal hyper-responsive towards the reduced number of, for example, marking posts, such that fouling of drinking water vessels is commonly encountered in caged animals.

Many social dominance studies involve feeding, on the assumption that the animal getting the small piece of food is the dominant one. Even if this is true, it certainly is not always based on animal size. I well remember receiving a gashed finger during a food preference trial with about thirty stray cats in one large pen. Small cubes of rat meat were offered to the cats and one small black and white cat of about 6lbs would flash in and out grabbing the meat by teeth or claws from any distance before the cats to whom the food was being offered had time to muse on the prospect. This was despite any apparent social groupings or status, because this cat was so focussed on the rat meat that it seemed totally oblivious of the other cats. There were no dominant or submissive posturings he just grabbed, swallowed and ran. In his singlemindedness he was equally oblivious of hooking my finger with a claw when swiping for the food.

In London's Fitzroy Square (see Appendix) when the small white matriarch was alive in 1977, at the time the feeding lady arrived and during her auxiliary feeding, if this cat moved towards a plate the other cats (most of which were about one third larger) made way for her, backing away from the plate. However, she was never seen to act aggressively and those retreating did so with apparent good grace. She was aunt, mother or grandmother to most of those present.

In the context of the feeding of close-knit urban free-living groups of cats food sharing is the normally encountered pattern and aggression is a rare event. Aggression appears to be far more common among the high density suburban house-cats than it does among the urban feral colonies of cats. It would seem that the parallels to the close confined state of the experimenter's laboratory are closest in the higher density living of suburban house cats.

It looks as if an aggression ranking only becomes noticeable with an increase in cat density. However, again in agreement with Leyhausen I find with penned cats, in high densities, that when they do show aggression, attacking or defensive movements perhaps over small food items, it will be very restrained. During such encounters, eyes quickly check the closeness of other interested cats and it seems that such close conditions inhibit a full display.

The early attempts to look for social dominance in the laboratory usually did not consider kinship. Despite my reservations on direct interpretation of laboratory behaviour studies, such studies do provide valuable insights into group structure if the context is remembered. The research work of the 1940s and 1950s generally showed aggression being most frequently expressed as frustration at being pipped at the food and not as a means of getting it. But even this was not unambiguous. The researchers could find a 'pecking order' of apparent dominance, however the top cat in any group did not appear necessarily to have a kingdom for life.

In the mid-1960s a further laboratory feeding-dominance study brought out the importance of the success of the first encounter between two cats in setting the pattern for a dominant/submissive role for future meetings. It emerged from this study that the animals, when living together in pens before the experiments were initiated, did not show dominance hierarchies until placed in the experimental setting and then not until after about an hour of such conditions. However in continuously group-living animals, such as monkeys, aggression is directly related to success in getting food, but this is clearly not so for a solitary hunter like the cat. When aggression did occur in penned test conditions very little retaliation occurred.

So studies from penned high density groups of non-related cats have shown a 'peck order' when hungry cats are presented with a tiny scrap of food. Such conditions can occur with newly formed cat groups of strays, but it is not long before family ties start showing.

David Macdonald and Peter Apps' Devon farm study noted more amicable encounters than aggressive ones between the related cats. Indeed the only very marked aggression occurred when the mother cat thought her new kitten threatened but that was only brief for the related adult females were soon accepted as 'babysitters'. Nonetheless aggression was shown to an 'outsider', a non-group-member.

Although the amounts of sociable contact varied between the individuals, and changed from time to time, it was a consistently frequent feature in contrast to rare spontaneous aggression which was only seen as one per cent of the interactions in the group.

When conditions do however approximate to the high density penned observations, then aggression can raise its head. In the mid-1970s John Laundré in the United States found that 98 per cent of aggressive encounters he recorded took place during the milk feeding sessions of his Wisconsin farm cats. The cats were given the milk towards the end of milking and as feeding

---

The amount of friendly contacts made between different cats of the farm cat group in David MacDonald and Peter Apps Devon study in two periods of the spring of 1978. The arrow direction shows which cat is approaching and which approached, while the thickness of the arrow (and the number beside it) indicate the percentage. During April all three queens were pregnant, and queen B in particular kept 'making up' to the tom. The kittens of queens A and B died of infection and following this those queens' friendly contact increased. In both months the cats remained very sociable. Overleaf: Over two months 72 aggressive encounters were noted in the Devon farm study. Half of these were between group members but were very mild, all of the serious aggression was directed by group members against an intruder tom.

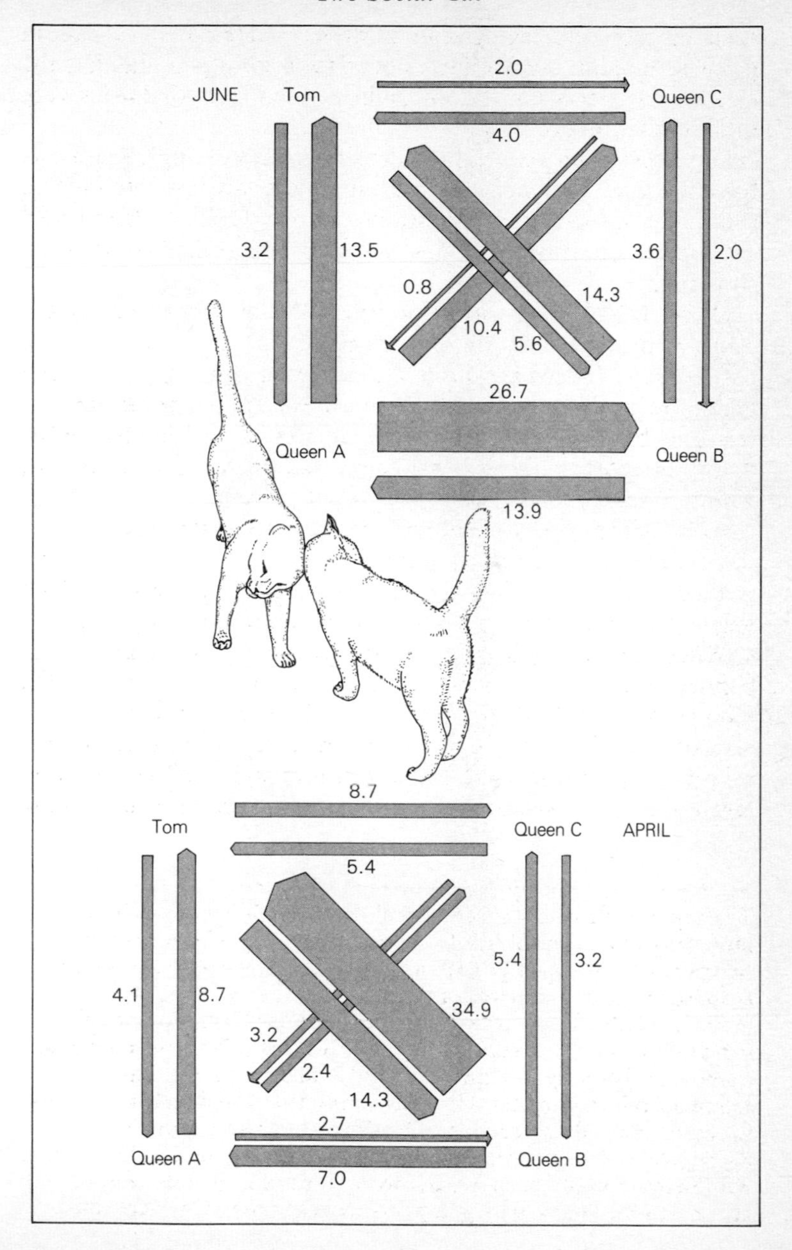
JUNE
Tom
Queen C
Queen A
Queen B
2.0
4.0
3.2
13.5
3.6
2.0
0.8
14.3
10.4
5.6
26.7
13.9
Tom
Queen C
APRIL
Queen A
Queen B
8.7
5.4
5.4
3.2
4.1
8.7
34.9
3.2
2.4
14.3
2.7
7.0

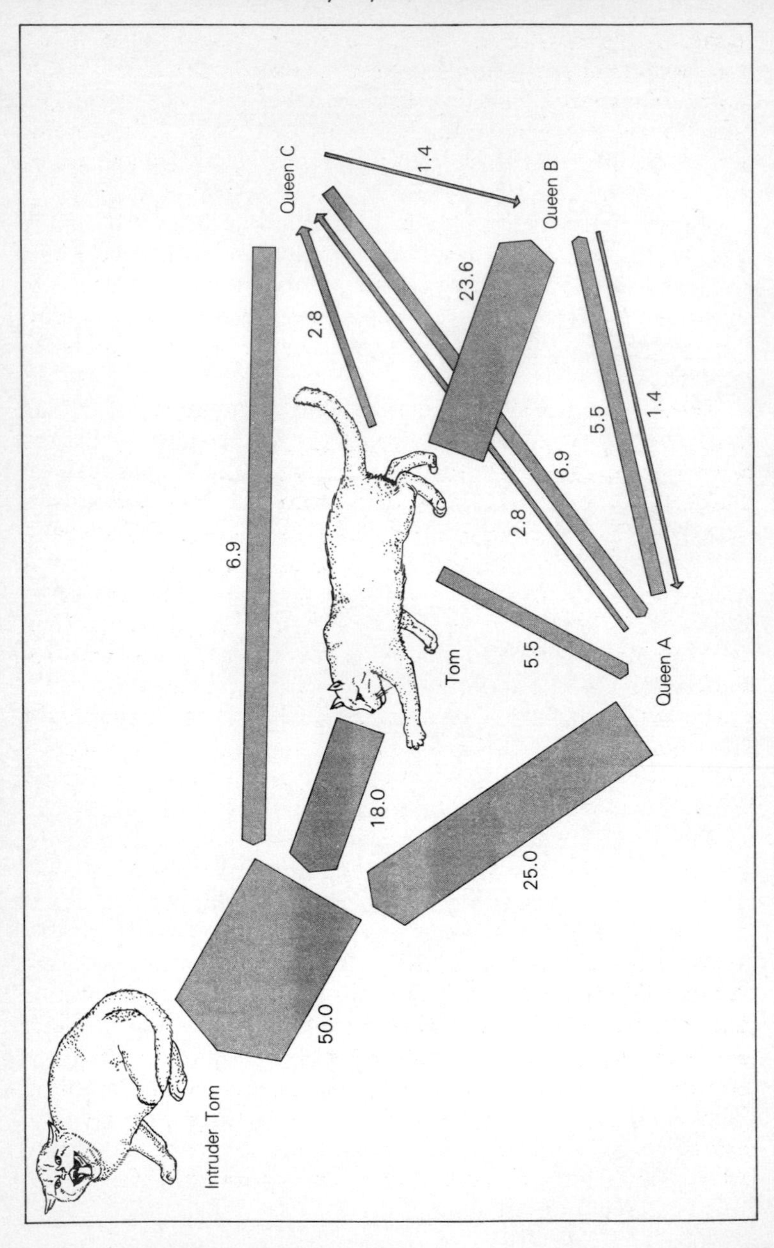
Intruder Tom
50.0
25.0
18.0
6.9
Tom
5.5
Queen A
2.8
6.9
5.5
1.4
23.6
2.8
Queen C
1.4
Queen B

approached an increase in tension arose and although when feeding started aggression stopped, the overcrowding led within half a minute to a flare up. Laundré noted far more aggression than amicable group bonding actions (4:1) during the feeding period, a direct contrast to the cats of Fitzroy Square where aggression was rarely seen. Why?

The aggression was so marked for the farm cats that Laundré was able to observe a female dominance hierarchy during feeding. However, when the group dispersed during the day encounters of any kind were so few between cats that social order was apparently meaningless. As the males were not aggressive he was unable to rank them.

The answer probably lies in the feeding conditions rather than farm lives making cats hyper-aggressive. Feeding at Fitzroy Square is not confined, with 3–5 dishes available the cats can eat their fill, while at the American farm milking area feeding was crowded around one dish and there was not enough to go round.

The minimal availability of food seems to invoke an otherwise latent hierarchy. Such a 'when the chips are down' pecking order may or may not have a role at other times. Certainly I find feral toms in many groups are often more tentative and wary of approaching food and arrive later in feeding than females.

The similarities of many features of behaviour between the Wisconsin farm cats and the central London auxiliary fed cats is striking in many features. For example the London Fitzroy Square's matriarch had her parallel in Wisconsin where the dominant cat at feeding time was also the oldest female of the group. A difference between the groups was that the Wisconsin cat maintained its position by aggression, while the London cat did not appear to, but merely 'queened it'!

As feeding time approached an increase in tension occurred among the Wisconsin farm cats. This also happens in the Fitzroy Square cats. At about an hour before the feeding lady is due to arrive the first one or two cats will take up position in the main core area. These cats will generally sit or lie quietly, for example in Fitzroy Square often by the roots of a tree. However they will maintain a personalised space up until the arrival of the feeding lady. Gradually the number of cats arriving in the main core area increases up until feeding time.

From about half an hour before she is due to arrive the tension can be seen to increase, but is not usually released by aggression in this type of colony of cat. Instead eyes and ears are trained on the route that the lady (frequently with small trolley/basket on wheels) will approach from. The tension can be particularly seen in Fitzroy Square by individual cats suddenly standing up and making short dashes in a fast trot to the basement railings of the surrounding houses. Equally spontaneously the dash is reversed, the cats re-entering the main core area or the attached extended waiting area and again sitting quietly by with a certain tenseness, eyes and ears trained. The distance from the square's railings to the houses' railings is most frequently accomplished in three brief spurts, being interrupted by pauses lasting a few seconds to several minutes, with the senses being freshly attuned to the auxiliary feeder's approach. (The extended waiting area is a bulge that appears on the main core area when the lady is likely to appear at any minute and if she is at all delayed).

When the auxiliary feeder arrives and feeding begins the personalised area is forfeited and side to side body contact and affectionate nose greetings occur between some cats. Feeding occurs quietly, with cats eating but stopping every so often to look anxiously about, apparently to a distance beyond the group.

Following satiation, individual cats become less tense and retreat to distances from 10–20 feet into the main core area from the feeding area. Here they relax and wash carefully. About half an hour after feeding they will begin to disperse, although some two-thirds of the cats are still likely to be in the main core area 1–1½ hours after feeding.

CHAPTER 6
The Twilight World of the Urban Cat

The urban feral cat lives in a quiet twilight world behind man's brash bright active centres. It is in the drab anonymous side alleyways that the cat meets the nocturnal wildlife of the fox and hedgehog. Particularly in the urban, rather than suburban, settings the site of the colony feral cat may on occasions coincide with the 'colony site' of London's down and outs. Although the gentleman of the road image may be one of romantic freedom in a rural setting, the urban situation of such unfortunate people is very different. They are sick men with drink ruined livers and kidneys, badly nourished and susceptible to passing infections. A winter of infections can take a high toll from such a group of people.

Occasionally a resentment against their co-habiting cats grows in the odd individual to the extent, as I have seen, of dispiritedly hurling a brick in their general direction. This is understandable for in such places as abandoned collapsing warehouses the mess and smell that tom cats can cause irritating enough to householders, here increases the general depression. However in the main a general tolerance exists between the men and the cats living in a nether world. Besides some of the men prefer cats to no cats and the probability of rats nearby.

Although I have found the income range of auxiliary feeders crosses the entire spectrum from film stars to hard-up pensioners, with the latter predominating, I have also come across an instance of down-and-outs with minimal income nonetheless buying cat food and feeding feral cats with it.

Although the urban climate offers some protection against the sharpest of frosts, nonetheless winter cold hits both feral cat colonies and groups of down and outs. Strangely the same tactic to keep warm is employed by both. Large cardboard boxes discarded outside small industrial premises (usually clothing manufacturers) are gathered up by the men and one box inverted inside another and maybe a dozen or so boxes may lie alongside a wall and only the breath condensing in the air above cracks in the boxes reveal occupancy. Every householder with a cat knows the obsession of domestic cats to sit in cardboard boxes. Colonies of feral cats will also sit in similar cardboard boxes, as long as they are dry, just like the men.

The urban climate

The urban cat is not so exposed to the elements as a rural cat in a similar geographical location. This protection is greater in a large conurbation such as London where the sheer mass of heat radiating from human habitation can raise the night time temperature of the central areas by as much as 10°C. The difference to the surrounding rural areas is more usually only a degree or two, but when this can lift the temperature above freezing this aids survival. During daytime the city area rarely exceeds one or two degrees above outlying districts due to the general warming effects of the sun.

Although man and his artifacts modify the local climate the local geography and topography still exert their influence. North London lifts to a series of ridges and hills, including the ridge through Epping Forest, and South London sits on the North Downs backslope. Such higher areas have lower temperatures and stronger winds and increased rain and snow.

The more central areas however have a real urban climate where continuous buildings reduce the mean windspeeds, but increase gustiness as anyone walking along Oxford Street past a side street will testify as they try to remove the grit from their eye. The buildings and a certain amount of airborne particles reduce the amount of ground incident sunshine, but I have never noticed that this has stopped an urban cat finding a sunny spot to bask in.

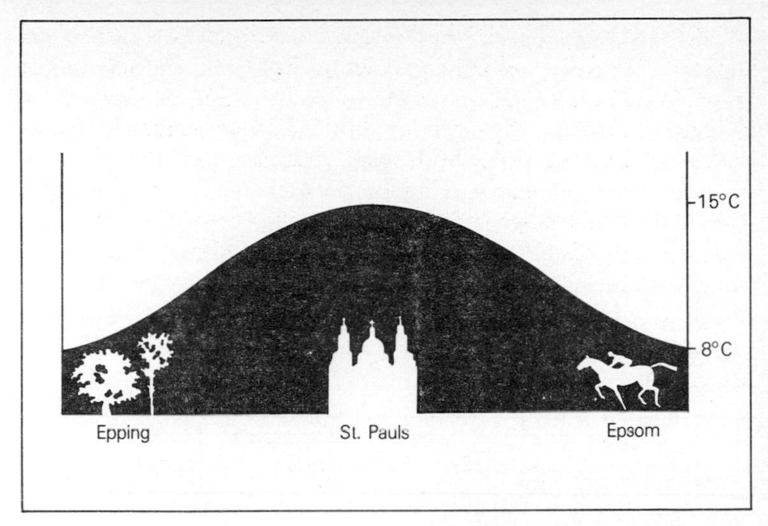

The bricks of London act together as a huge storage radiator, not allowing the nightime urban temperature to drop as much as outside beyond suburbia.

Distribution of urban man and cat

London, like many cities, has been built in distinct waves of development. For much of its life its inhabitants were housed within the area contained by the city walls and this was so even up to about the time of Samuel Pepys, when the size of overspill was still slight. This is in complete contrast to today, when the lowest population levels are to be found in the city to the extent that it can be considered virtually uninhabited, a bustling, busy and noisy place that no-one lives in. London's development, as for many conurbations, has left it a bit like an onion, the zones clearly identifiable as you drive out from the centre. Like all organisms its spread has followed a normal growth curve, small and contained for a considerable period, to be followed by an exponential increase, which we are still in. The first big wave in the mid to late 1700s was the selling of large estates whose names are still commemorated for instance in street names of West 1. This was building in a grand manner for the architectural feel of railinged terraces around squares and crescents (the study area of Fitzroy Square is in this zone). Mews fill any potential individual garden area, giving only the pattern of close streets, with alleys and the occasional formal communal landscaped garden. Such urban settings ideally suit the colony formation of stray and feral cats. Domestic cats find it particularly easy to demark on a territorial basis farther out in the suburban house and garden with group identification to a human family. This is far less practicable for the eighteenth century terraced housing when it is an amalgam of offices and multiple tenancy. With such flats packed on top of each other, with only a communal square which the human family rarely if ever visits, the domestic cat not only lacks a definable area but has great difficulty in forming a territory and home range directly comparable to the suburban cat.

The ease with which feral cat colonies hold such areas means that individual domestic cats have to trace their own movements amongst the colony delineations. This is the exact reverse of the suburban setting where strays have to move among the tight confines of domestic territories and ranges. There are breaks in this urban setting and blocks of flats may have a larger surrounding garden, but the absence of human delineation and identity in the garden causes greater feral/domestic contact and the domestic cats often form a grouping among themselves similar to a feral colony. In one block of flats in particular where such a hybrid colony exists in South London, an old metal bin, about the size of a dustbin, long abandoned among some trees, has become semi-covered over the years and is used by nursing mothers to have and raise their kittens. The bin is just outside of a core area for the hybrid colony cats.

With the advent of greater public mobility, railway and omnibuses this urban style was superseded by the building of inner suburban terraced houses with small gardens and then further out with slightly larger gardens from the mid 1800s. Much of the terraced study area of Barking (where sizes of domestic cat home ranges were looked at) was developed from 1890–1910 and can be seen as mid-suburbia. Outer suburbia with the advent of the between the wars semi-detached house is typified by the north Romford study area. In terms of garden to house ratio, post second world war housing can be seen as an adjunct to the between the wars outer suburbia. Generally the further from the 'onion's centre' the more garden space.

Both mid and outer suburbia, from the domestic cats point of view, are very similar. Mid-suburbia may hold a higher density of domestic cat numbers due to closer housing, but fundamentally the human population clearly delineates land on which a house is sited as a kind of territory. More importantly human movements within that delineation are recognisable visually as much to the cat as to other people in terms of objects associated with a family and in terms of scent and marking activities.

The suburban cat pattern is seen repeated in many small towns and villages, due to the similarity of the housing/garden patterns.

In studying cats we tend to focus on the significance of such actions as scratching posts as visual signs to other cats. It is easy to overlook the significance of our putting in plants and cutting grass as territorial claims among the human population. It is these same factors, plus our presence during these marking activities that reassures our domestic cat that this is 'our group's' core area.

The lack of human delineation and marking seems to allow high density cats to form colony associations in urban conditions. In contrast the clear delineation and land territoriality of suburbia still allows a colony identity for high density cats, but in this instance we are the conspecifics and this seems to give the appearance of solitary cats. In rural settings cats may appear more truly solitary but nonetheless groupings or colonies focus in such places as farms, under hospitals, factories, in a similar manner to such focal settings among suburbia. This preference in the cat for association does not make the cat much less of a loner in outlook. When wandering or hunting about the home range or its adjuncts, the cat is very much the solitary animal of belief and folklore. Kipling's cat that walked by himself still does so, even if now we see a partly social animal.

London's high cat density arc

The high cat density area to the west of the city in London corresponds to both availability of food and shelter. The availability of food comes from ash-free plastic refuse sacks and more so from auxiliary feeders, both dedicated elderly ladies and to a

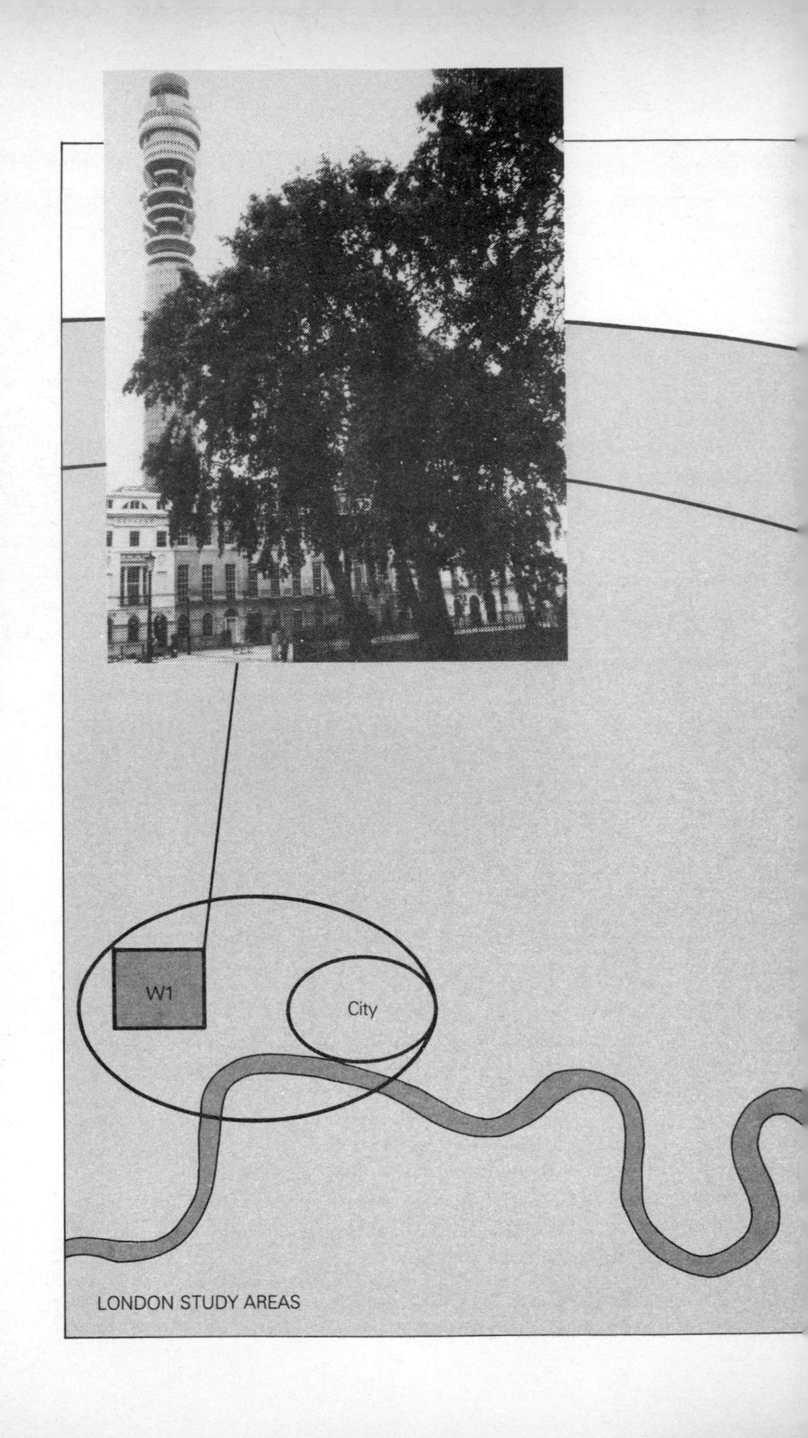
W1
City
LONDON STUDY AREAS

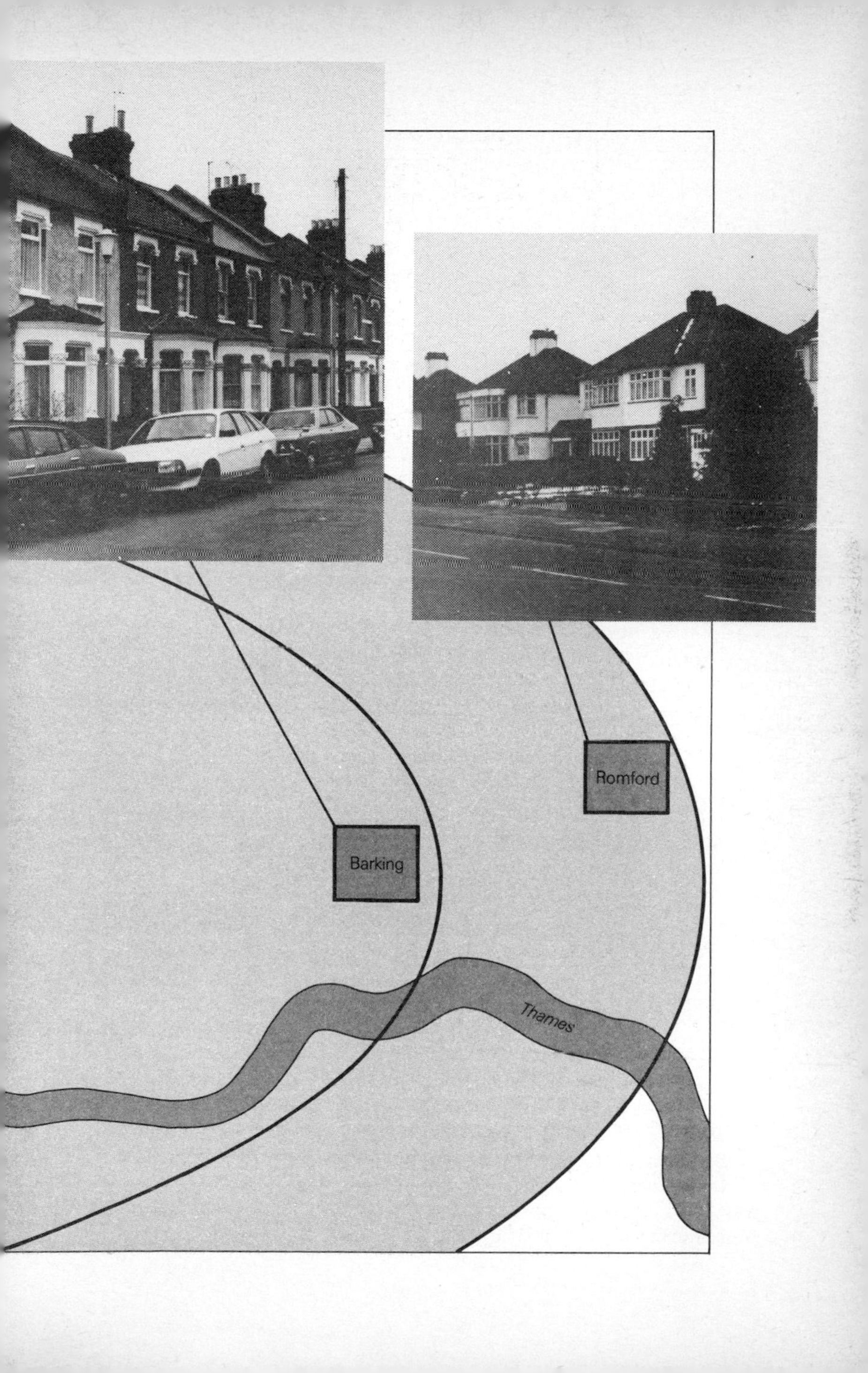
Romford
Barking
Thames

lesser extent office workers. Soho appears to have a slightly lower density of cats than the main high density arc and yet has a greater availability of refuse from restaurants. This seems to imply the auxiliary feeding may be more significant with regard to cat numbers, but feeding is not all. Shelter is of particular importance and the map of the high cat density areas concurs closely with those areas with railings and basements. The railings provide protection from disturbance even when the railings are wide enough to admit dogs, which being longer, lose the agility of a cat in passing through railings. Fitzroy Square colony is regularly met by the local vicar's dog when out for a walk of an evening. The dog will bark at langourous cats on the other side of the railings until it cannot stand the frustration any longer and rushes easily through. However, not as easily as the cats rush out. The scene is repeated a number of times, then the dog continues its walk. It seems a fairly amicable arrangement on both sides.

Railings do not just provide escape routes (which they do utilise to the full) and access to basement shelter, but secure areas within which the cats can go to sleep in the sun without fear of being trodden on. The concrete spanning bars well out of reach of both basement and pavement are most frequently used for solitary daytime snoozes, but some cats prefer to sleep near a conspecific cat.

London was still a fairly compact town in 1750, but over the next hundred years a number of estate lots were sold and built upon. Designers/builders such as the Adam brothers put up terraces in much of the present high cat density area in this period. It is this historical quirk that has grouped much of London's railings in one area.

Although the above features of feeding and shelter (allied to low disturbance) seem major parameters for the area's high propor-

---

Map of the recorded distribution of feral cat colonies in central London for the year 1978–9 (153 colonies). As well as the Thames, the position of major parks are shown for reference points. The high density arc runs SE from Regent's Park towards the Thames. The map cannot be an absolute map of all colonies, but it is probable that the pattern is fairly representative.

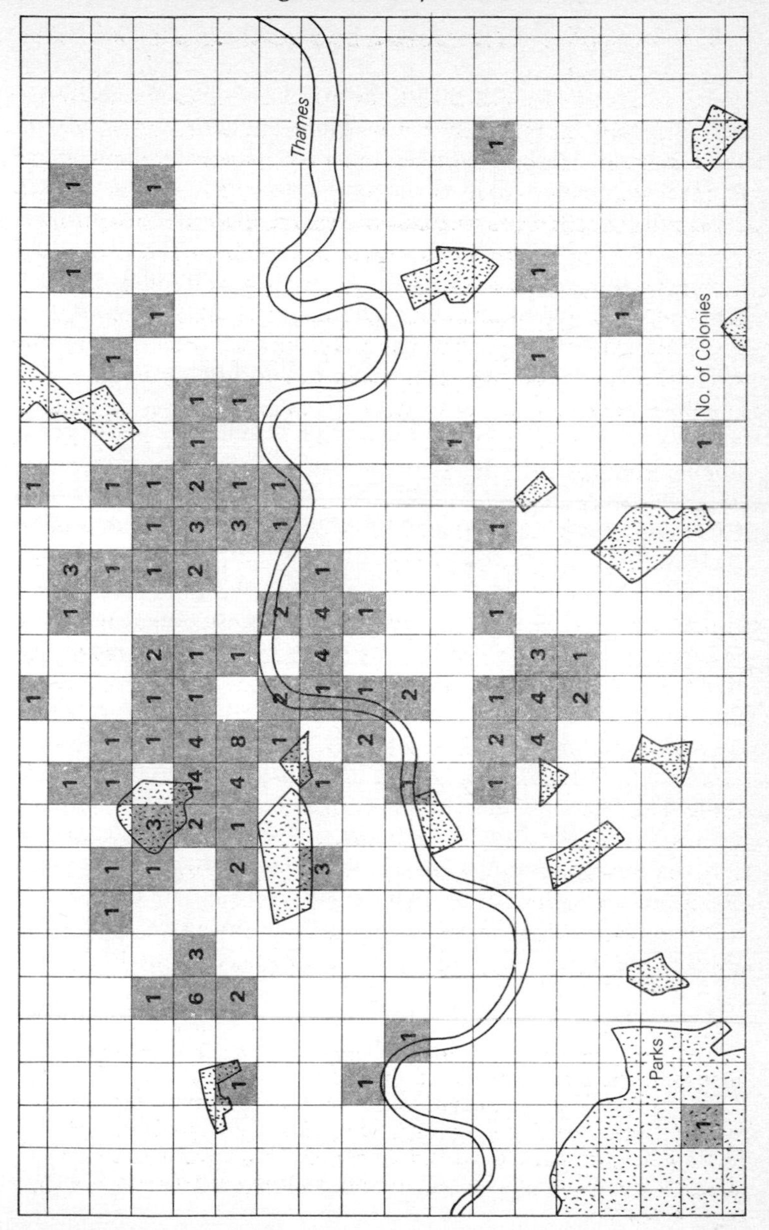
Thames
No. of Colonies
Parks

tion of cats, it is possible that they are re-inforced by certain other factors.

The area is, perhaps strangely, very highly populated with overall range of between 40 and 168 persons per acre (based on the 1961 Census). This compares with the Barking study (Chapter 5) area which has a range of 12.5. to twenty-six persons per acre and a rural population beyond the suburban limits of 0.1.–1.17. persons per acre (but within the thirty-five mile range around London). This allows for a higher probability of auxiliary feeders and of a domestic cat pool that could stray. The urban depopulation that has occurred is primarily within the city and its immediate environs, while the areas such as West One have very high populations.

More significantly there is a higher percentage of elderly people in the area than for most of central london. Furthermore immediately to the west of Regents Park down to the Thames there is a very high percentage range of 20–80 per cent of the population at pensionable age. This may provide nothing more than a potential pool of dedicated auxiliary feeders, but it does mean a small number of children in the area. Cats do not always see eye to eye with children over matters of peace and quiet and sleep. (The range for the area of children under five as percentage of the population is 0–4 per cent and of under fifteens is 0–14 per cent.)

I compiled a distribution map of recorded feral cat colonies for central London for the year 1978–9 from a combination of sources. One was from the records of a specialist pest control company, another from the records of an active cat charity, also from the response of readers to an appeal in the *Evening Standard* newspaper, and from my own accumulated observations. Like most distribution surveys this map cannot pretend to give a full picture. There will have been many feral cat colonies unrecorded on this map for that year. Nonetheless as the methods of obtaining the data were sufficiently different and as they basically re-inforced each other it is quite likely that the *pattern* of colonies may be fairly representative.

Certainly the north-east quarter of London's W1, on either side of Great Portland Street, must have the highest density of feral cat colonies anywhere in Britain. As long-term established colonies they are, of course, identical to T. S. Eliot's 'Jellicle Cats' being mainly black and white.

Suburban cats

In my north Romford survey of over one and a half thousand households carried out in 1977 at an adult education college there appeared to be little correlation between subject matter studied at the evening classes and whether the household represented kept a cat. This can hardly be expected, for even if a 'certain type' of person is a 'cat person' their choice of an evening class cannot be taken as a full reflection of such a disposition and even if it could the disposition of one person cannot be read as the wishes of an entire household. Nonetheless I did find it interesting that in 'Men's Keep Fit', a class representing 28 households, no cats were kept, while in a 'Creative Writing' class 100 per cent of the households kept cats!

I carried out the survey in north Romford because the area is typical of much of outer London suburbia (and in consequence most typical of the country's housing overall). Much of the housing was built between the two world wars and is in uninterrupted development from central London. London is not as circular about the Thames as it might at first appear, but sits higher above the river and due to the rapid expansion of East London from 1840–1920 took on a bulge in an eastern direction. This bulge continued outwards, engulfing such towns as Romford, so that much of north Romford is in the continuum of the outer London suburban belt. The 1556 households contained 501 domestic cats, of which 109 had previously been strays, that is 22 per cent of this suburban cat population had been adopted from the ambient pool of strays. The number of households with cats was 436, giving 26 per cent of all the households actually 'owning' cats.

The difference between the number of households with cats and the number for total cats is due to the instances of more than one cat being in some households. The pattern for this can be seen where a clear majority can be seen to live with one cat to a household and only just above 20 per cent of the cat households

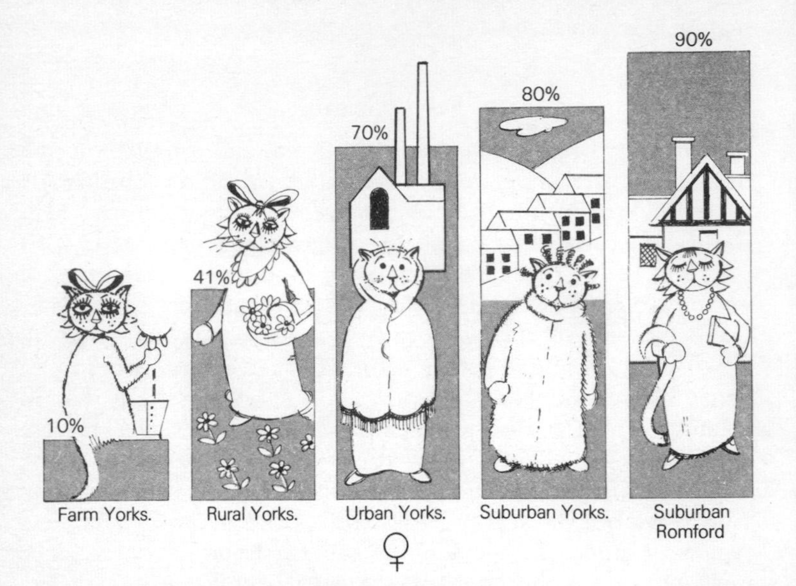

Very few Yorkshire farm cats were found to have been neutered (Recent Yorkshire survey by Colin Howes), and the figure was still fairly low for rural house cats. Towns and cities have a higher percentage of neuters, where the greater density of cats makes the associated effects of intact cats more telling! Suburban Yorkshire and suburban London (Romford: author's survey), with their higher densities of cats, have the highest proportion of neuters.

(or one in five cat households) had two cats in residence, while under 1 per cent of cat households had more than two cats.

With regard to the cat population size, I found 86 per cent of the cats had been neutered, leaving only 14 per cent of the domestic cats intact. As there were more females than males it was not surprising that a greater number of females were neutered than males. However, even when allowance is made for this, a slightly greater proportion of females were neutered than males.

From the neutering figures it would seem that suburban householders are currently taking definite steps which must result in less breeding of an unrestricted nature. Certainly both economic and social pressure exist to make this expedient in the form of

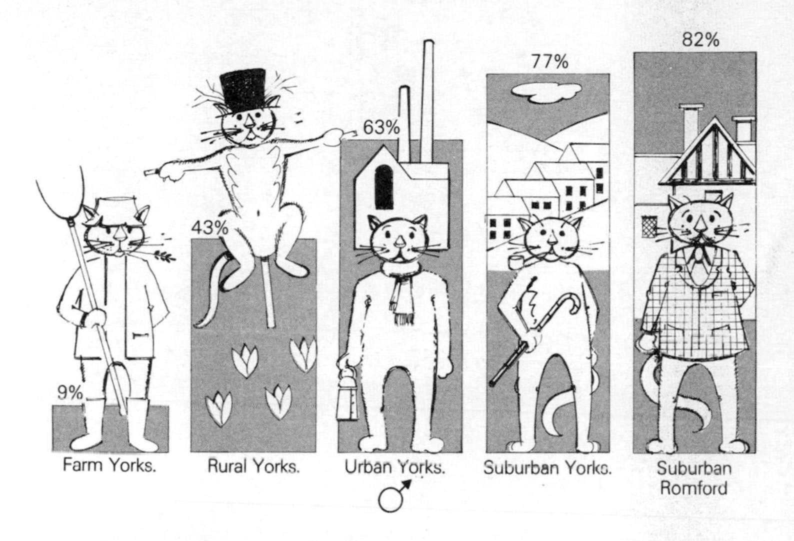

unwanted kittens which require great effort to home, and the strain 'on the nerves' of both suffering the regular calling of an oestrous queen, and of the all-invading pungency of the smell of a male intact cat. In towns where there is a shortage of available kittens, such as Cheltenham and York, such expediency has sufficiently changed the state of affairs through domestic neutering, that cats are genuinely wanted, rather than just dumped on unsuspecting friends.

When widespread neutering in the Romford population is supplemented by the large figure removed from the stray or feral state (of over one cat in five of the domestic cats found to be so adopted) it would seem that the state of influx into the domestic population of free-living stock is considerable in suburban cats. This has great relevance for the breeding and genetic strength of a population. In terms of cat population control, with so much occurring without outside help, slight help with either factor would no doubt influence the free-living population level. Such a rate of turnover may also be contributory (along with high domestic cat population densities) in limiting suburban feral cat colonies to breaks in the suburban housing pattern.

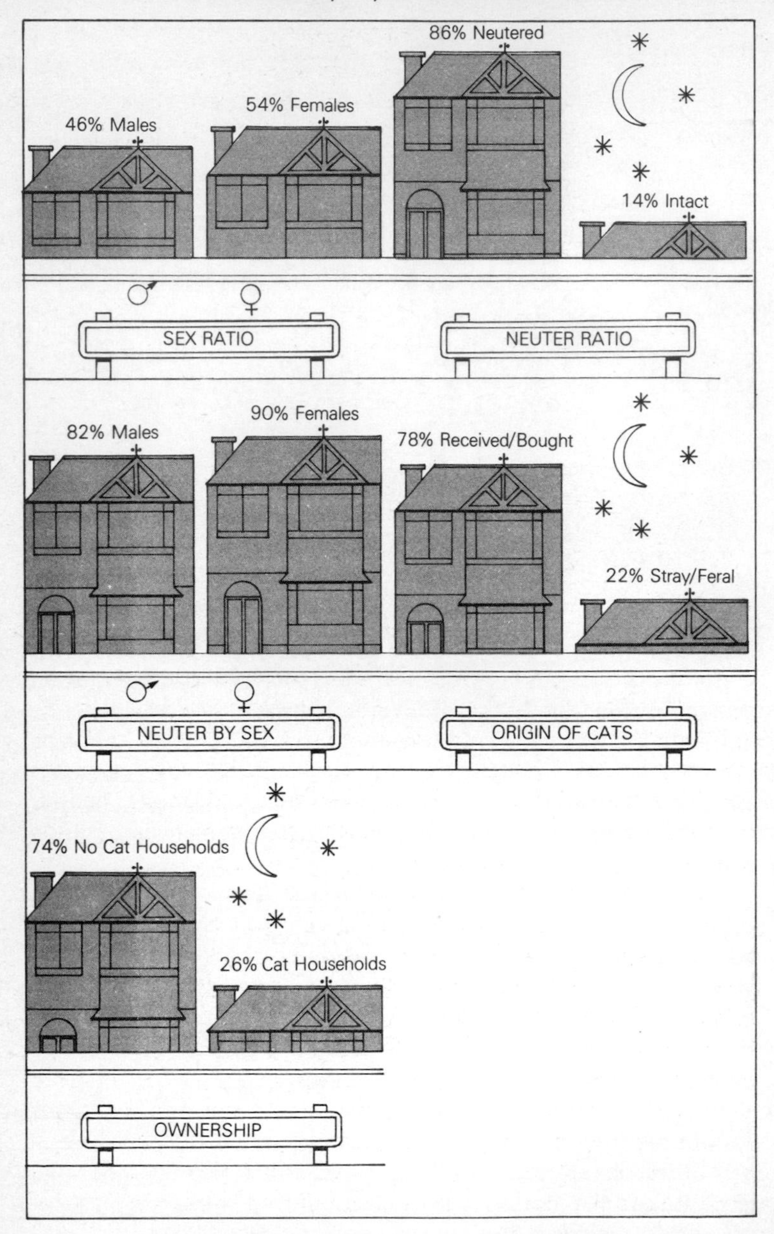
46% Males
54% Females
86% Neutered
14% Intact
SEX RATIO
NEUTER RATIO
82% Males
90% Females
78% Received/Bought
22% Stray/Feral
NEUTER BY SEX
ORIGIN OF CATS
74% No Cat Households
26% Cat Households
OWNERSHIP

Other surveys

The Romford survey was in an area of relatively new suburbia and the 26 per cent households with house-cats is consistent with previous surveys for the same general standard of housing in Britain and America:

		per cent households
1944	Newport, Wales	26
1944	Cardiff, Wales	30
1974	California, USA	25
1975a	Glasgow, Scotland	25
1975b	Glasgow, Scotland	39
1975	Massachusetts, USA	24
1978	Romford, England	26
1979	Doncaster, England	29
Mean figure		28

However unlike most of the urban/suburban surveys, one rural study in California, USA has shown that keeping cats is easier and perhaps needed in the country, for 45 per cent of the households had house-cats. Similarly a study of rural Yorkshire in England, found 61 per cent households with house-cats and 94 per cent of farms with farm cats, over a thousand cats living on just over 200 farms.

London outer suburbia (Romford) survey.

Each roadsign refers to the grouping of two 'houses' behind it. The height of each suburban house represents the percentage of cats in a particular category (further shown as the number above the house). For example, under 'ownership' about a quarter of the households (26%) had a cat, while about three-quarters (74%) did not.

CHAPTER 7
What Does a Cat Eat?

Many observers of the domestic cat, will say that its life seems to be one largely of eating and sleeping, sleeping and eating! But when the cat has to support itself, what does it eat then?

What does the self-supporting cat choose?

Some animals are very inflexible in their diet, but this does not seem to be the case for the cat, even when returned to the wild. I say this despite everyone knowing that 'their cat' is very fussy and will eat only 'such and such' cat food, or only chicken or whatever, to the extent that the cat will put its foot down and not eat even for some days until its wishes are granted. This is really the cat showing a preference and being a strong-willed animal, withdrawing one of the few things we will recognise until we have learnt our lesson. It is playing the maternal game and continuing our training.

Cats living wild also express preferences for certain bait foods over others. Roast chicken and pilchards are both considered highly, as is also the rabbit flavour of one particular brand of commercially available cat food. Whatever the promotional television adverts may say, cats when left to their own devices, unhindered in the wild, vote with their feet for one particular tin! Strangely enough although I have not eaten the cat foods myself, a duke I met in the course of writing this book, who was brought up in the grand manner with cheetahs on his father's estate in Italy, has such high regard for his cats' stomachs that in consequence has tried all varieties of commercial tinned catfood. He independently had come to the same conclusion about this particular variety of cat food (commenting in passing that some of the rivals were unspeakably foul) but that it was at its best – fried!

Preferences aside, living in the wild one can be adaptable or if your particular food dries up – starve. Cats are adaptable.

Rodents were found to be the favoured prey, being three-quarters of the diet of isolated forest feral cats in Maryland, USA.

In a study in Oklahoma, USA, urban feral cats were found to catch and eat a smaller amount of animals and birds than garbage which they scavenged. This was also found to be true for Pennsylvania, USA, but the animals tended to catch less prey and depend more on garbage (in both town and country) in winter and autumn than they did in spring and summer.

The cats looked at in a study in Victoria, Australia, in the early 1970s showed themselves to be living a life almost as independent from man for their food as it is possible to be. The cats were found miles from any buildings in open bushy or agricultural land. Virtually all the food they had eaten was caught (or found as carrion), only a trace being garbage for which few animals had considered it worthwhile making a very long trek. However, not only is the feral cat an introduced animal into Australia's unique marsupial wildlife, but so are the mouse and rabbit. The cats in the bush were found to be living off both native marsupials and rabbits and mice. However it is worth wondering what Australia's wildlife is in most danger from, the hunting feral cat or agricultural man. On the agricultural land rabbits and mice dominated the cats' diet, while not one indigenous mammal was found in a cat. Both cats and native mammals are in existence in the bush which is more than can be said for the native mammals on the agricultural land. The cats are preying on the available food. In the bush they have slipped into the vacant niche of the native 'cat' (Dasyurus viverinus) which is now very rare.

With the marsupial 'native' cats and introduced feral cats the situation seems very much 'did he fall or was he pushed'. Has one cat directly confronted the other, has one cat caught foreign infections from the other or did one decline independently anyway and the other fill its place. All the same arguments that we are familiar with in England for the red and introduced grey squirrel tussle. I think that although there may be some truth in each explanation that most leverage may lie in the very similarity of such animals. Both compete for the same food, for the same 'ecological niches', but due to their similarities of manner they could almost be treating each other as conspecifics, and allowing territoriality to each other. But as one is slightly more 'efficient' in some way (perhaps less specific in diet) such that it is either more long lived or prolific, with the demise of a native animal the nearby territories of the introduced animals engulf the vacuum.

The pattern in America is repeated in England in as much as the rural cats caught more than the town cats. However, this was inevitable because the volume of caught food in the urban cats examined was negligible. Food for which the cats had foraged (garbage and vegetation) was the same in both cases. Slightly less bait and auxiliary food was found in the country cats, but it is in this section that a huge difference to the American and Australian cat lies. These are free-living cats and wild (sometimes very wild) and yet nearly every colony of free-living cats I have encountered in England has at least one lady (usually elderly) feeding it. Sometimes it is a man, (but rarely,) and some colonies have a number of secondary auxiliary feeders in that on their way to work some people fling in a bit of food (sometimes regularly, sometimes erratically). The primary auxiliary feeder, true to type, will be selflessly spending a large proportion of her income on food for the cats. As the traps were baited by cat trappers with similar food to that used by the ladies it has not been possible to differentiate between the bait and auxiliary feeding. However on occasions where the food types have been sufficiently different, it appears that the amount necessary to trap the animals is not large and thus it would seem that the staple diet of England's feral cats is 'delivered to their saucer'! This certainly ensures the full-bodied look of our gone-wild cats compared to some of those around the Mediterranean.

Origin of food consumed by feral cats in America (*above right*) and England (*below right*). The American study by Eberhard in Pennsylvania found rural cats eat more caught prey as food than urban cats, and the same was certainly found by the author in the UK. Unlike the earlier American study it was possible with the UK study to distinguish between garbage and auxiliary feeding.

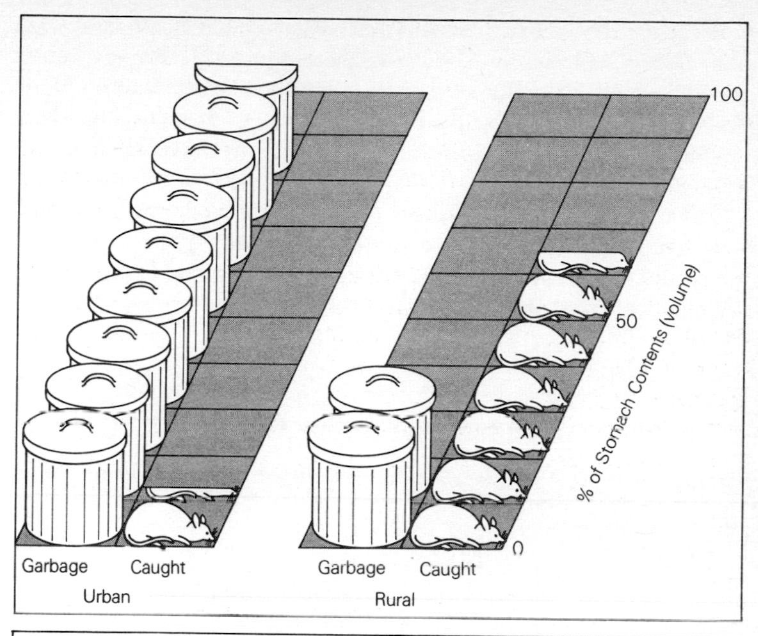
100
50
0
% of Stomach Contents (volume)
Garbage
Caught
Urban
Garbage
Caught
Rural

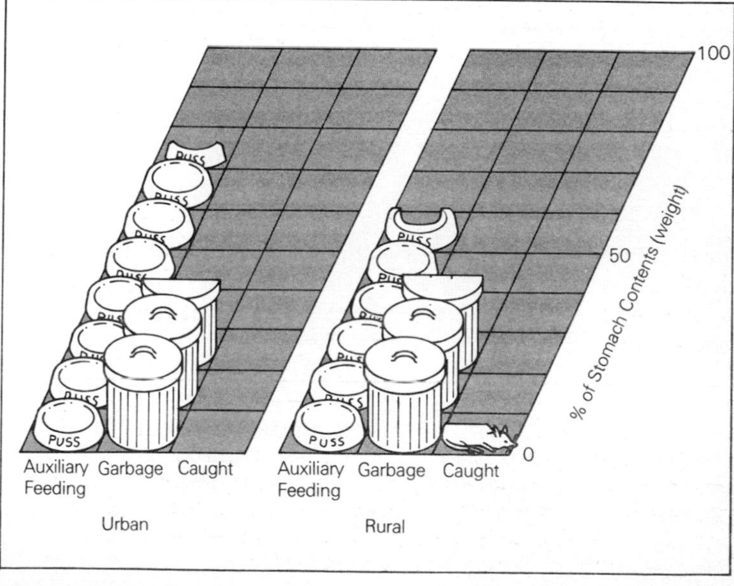
100
50
0
% of Stomach Contents (weight)
PUSS
Auxiliary Feeding
Garbage
Caught
Urban
Auxiliary Feeding
Garbage
Caught
Rural

Perhaps they are harder-hearted when it comes to feral and stray cats in America for the Pennsylvania and Oklahoma studies showed no auxiliary feeding. Alternatively as those studies were undertaken in the 1940s and 1950s, when tinned cat food was less universally used, some of the 'garbage' may have been kind-hearted auxiliary feeding.

When vegetation is looked at from foraged material and scavenged bin material, more scavenging occurs in the town than the country. Nonetheless with auxiliary feeding and foraging our English cats don't seem to be catching too much!

Hairballs and other inedibles

Cats in the wild state, due to hair swallowed from grooming and hair from prey, regularly vomit the hair ball free as a cylindrical mass, usually at 7–14 day intervals. Nonetheless some hair will pass beyond the stomach into the small intestine. Normally it will pass through the intestine rarely causing further trouble.

In domestic cats this regularity can be lost and with an uncleared accumulation of hair matted in the stomach, understandably the appetite can be lost. In cases like this the vet is still likely to reach for the old remedy – liquid paraffin. Needless to say, if the cat is moulting heavily, some extra grooming with a comb will reduce the number of times impaction occurs in cats that are prone to this condition.

Unlike in the dog, with its habit of retrieving objects thrown for and to it, the domestic cat is rarely presented to the vet as having swallowed inedible objects. In the feral cats however I find the stomach often has up to a gram or so of grit in it. The grit is mostly in the range from the size of grains of sand up to that of a pea. I have also found that they have eaten pieces of wood, bits of match-boxes, and even a small 2g. metal box. The box had evidently been in the cats stomach for some time, for despite being nearly sealed it was crammed full of hair. It might be wondered why a cat should swallow a metal box and where it would find a swallowable box. The last part at least I can answer – underneath one of central London's large teaching hospitals.

Plant-eating cats

Cats, being carnivorous, do not have an arrangement of teeth that can deal with vegetation and yet they regularly can be seen sitting beside a clump of grass, awkwardly trying to chew off a blade, trying to shear it with their side teeth, after which they gulp it down. Unlike herbivores such as sheep (or even omnivorous animals like us) cats have very little lateral movement in their jaws with which to grind the grass.

So why do they eat grass and how much and when? For that matter, is grass the only vegetation they eat?

Many cat owners will say that a cat eats grass when it is not feeling well – as medicine. There may be some truth in this in that it may be eaten in response to stomach discomfort. However, as anyone soon discovers who makes a point of following their pet's activities throughout the day, day after day, cats regularly eat grass, in most pets at least each day if not deprived of access to the outside world.

The reason for eating grass may be as apparently simple as providing 'roughage' in the diet to aid the passage of food through the intestine. The cat equivalent of the bran cult? The cellulose cell wall structure of grass (or similar vegetation material) being hardly digested by the gut action could help to bulk and structure the bolus of food passing through the system.

Unlike us, cats can synthesise vitamin C, so their need for vegetation should not at first sight be the need to supply this vitamin. However they may need to supplement this.

I have found in English feral cats that just under 10 per cent of the stomach contents consists of vegetable material. In country cats the amount is just a bit less than in town cats, but still very similar.

However this plant material is far from being all grass. I looked at the origin of the material as being either scavenged (such as household scraps from dustbins) or free foraged plant material (such as grass, hawthorn leaves etc). In the country colonies of feral cats the freely foraged plant material was just under 20 per cent of all the plant material, while in the town feral cats this was 6 per cent. So while the amount of plant material eaten was about the same, the town cats with less access to naturally growing vegetation compensated by bringing up the amount by eating vegetable material from garbage. (Of course it is possible that having eaten plant material from the garbage they felt less need to search for naturally growing vegetation).

Whatever the motivation, it is interesting that both sets of free-living feral cats felt obliged to eat fairly similar quantities of plant material. It might be worth bearing this in mind when feeding domestic animals that have no access to the outside world, such as in catteries. (Any plant-derived meal homogenised into the auxiliary cat food by manufacturers has not been allowed for here, and it would seem not by the cats either. It would perhaps be sensible therefore to have some plant material at catteries recognisable to the cats, rather than trying to hide it all.) Cats kept confined in flats, even if given a flower pot of growing grass, have a nibble at exotic potted plants. Bearing in mind some of these are poisonous and have been responsible for poisoning cats, augmenting the grass-pot with some plant material with their food may lessen this.

The unvegetarian cat

If the cat, when left to its own devices, eats about 10 per cent of its diet as plant material, could it exist on a higher amount? Some suggestions are sometimes given by vegetarian owners that their pets should adhere to their owners usual stance despite the cats carnivorous disposition. It could be argued that it is as immoral to impose on an animal with the body design and requirements of the carnivore a vegetarian diet as it is possibly moral to decide individually to turn vegetarian an omnivorous (and dietetically more adaptable) human. However, the argument does not turn on morality but on whether the cat can flourish on such a diet.

Kittens have been found not to gain weight on an exclusively vegetable diet (not even with vitamin B12 added with other supplements), but normal growth rates can occur in weaned kittens with up to a third carefully balanced vegetable matter.

Cat as scavenger

The role of cat as refuse scavenger must have seen some changes as the years have passed. In 1875 in Britain a Public Health Act was passed that determined that refuse should be taken from premises on certain days. Before that pickings must have been even richer, because from the scavengers viewpoint as dustbins are being filled the rubbish tends to obscure the lower layers. Before bins there was no problem, waste food (and other waste materials) were flung into the streets and these scraped into huge heaps. The coming of the paper and plastic sacks have made disposal quieter. However these are obviously more vulnerable to a determined claw than a metal bin, and the one time that the claw gains determination is when the carcase of a chicken is thrown away. The grease from the remains soaks through the sack and the passing tom sniffs – and forgets about passing.

The innovation of paper and plastic sacks has only been possible due to another 'revolutionary' change in our habits that is reflected in the changing composition of the refuse. As fewer people have open fires the proportion of ash in our bins has become less and therefore our bins less heavy. This means that the paper and plastic sack era holds more to the cat than easy access – it means clean food, not caked and smothered by gritty and possibly hot ashes!

The main changes that seemed to have occurred in Britain's household refuse over the years since before the last World War is that as the amount of ash has become less, paper has greatly increased, mainly due to packaging of goods.

These factors may have changed the availability of scavenged food to the cat and other bin scavengers, but the amount of edible material has remained nearly constant in weight (over 2kg or 4⅓lb per household per week).

Some of the increasing volume of paper (and for that matter the smaller volume of plastic) is ending up inside the foraging cats. Whether or not intentionally as a non-digestible 'roughage' fraction is open to debate, but the plastic certainly evades digestion, while the cellulose of the paper is resistant to digestion, but with time the paper will fragment. A certain parallel to the grass fibre can be drawn, for both paper and plastic film and cellophane wrappers end up in the same position in the cat's stomach as a grass impregnated hair ball. I feel that the cat eats paper that has become impregnated with fat or similar tasty material from other rubbish, taking it to be edible from the flavour.

In one of the preferred locations by feral cats – hospitals, there are two distinct forms of waste material. Infected dressings, human tissues and similarly contaminated materials are not available to cats (the usual method is on-site incineration). However about half of hospital waste is of the normal domestic variety, including kitchen wastes. Although some use is made of plastic sacks, most London hospitals use cylindrical bulk refuse containers (mounted on wheels) that each hold the equivalent of ten household dustbins. As half a dozen or so of these are usually grouped together in a quiet courtyard of a hospital and these bulk bins are uncovered, the cats are able to jump in and spend a considerable time unchallenged turning over the contents. The height of the wheels make these ideal shelters under which the cats can often be found resting.

Although the average weight of household refuse per week for Britain is about 13kg (30lbs) the character of the pickings for cats varies from area to area. In a brief survey of six streets in different parts of Britain thrown away food ranged from a carbohydrate excess of bread in Glasgow through to a pound of raw mince among other choice goodies being abandoned in Manchester. I have ceased to be surprised at the edible material thrown away that cats can scavenge. I was following the activities of a known feral cat through one of London's West End squares when I saw it leap up onto the rim of a street litterbin, teeter for a second or two and then dive in. I followed quickly on his heels, and peered in, unfortunately disturbing him from his prize. This was a raw, whole, fillet of fish, laid in solitary splendour in the bin. I am still speculating as to who would throw away whole fillets of fish in the street! Needless to say I left it for the cat to return to.

CHAPTER 8
Cat as Hunter

The kill

The cat family in general are efficient killers. Initial contact is made through the powerful front limbs with the protractile claws. While a number of the cats may kill by restricting the throat and suffocating large prey, most cats kill by an upper neck bite. This may cleanly separate the neck vertebrae or even break in the relatively thin bone at the back of the skull. This is the normal tactic of the domestic cat with a mouse. This ability to deliver a precise death stroke is in accord with an animal that, as a solitary hunter, has to overpower its prey quickly. Although the lion may hunt as a social group, even then the kill is normally carried out by one animal. For a pack animal such as the hunting dog there is no need to develop an efficient killing blow aimed at a specific part of the prey, indeed the animals would be biting each other if they did so and therefore the prey animal is brought down by attachment to a number of convenient places to hold. To the outside observer the canid method of killing looks messier and the felid method cleaner and better, but the method seems really to be derived from the differing social structures, which in turn derives often from habitat. It is perhaps due to the potential swiftness with which a cat can deliver a death bite that many people find the play activity often indulged in with prey so distasteful.

The young hunter

At about five weeks of age farm raised kittens in America had begun to eat food in the nest that the mother had hunted and killed. Following this initiation to solid food the kittens being studied began to follow the mother cat on her hunting trips. Also at this time the play of the kittens took on the movements of stalking and biting as of the prey, but in an erratic manner. Co-ordination and correct sequence of movement develop with the experience gained accompanying the mother's hunting trips. Kittens brought up without this experience are less predatory than those that have watched their mothers perform a kill.

In such behaviour that is considered 'innate' by most ethologists Paul Leyhausen stresses that ethologists do not consider that because something is innate it denies a 'developmental history', in other words something instinctive can still be improved upon by experience. Further, Leyhausen found that if the mother does not bring live prey (although in the early stages she takes care to bring only passive prey animals or ones that she has slightly damaged) and present it to the kitten between their sixth and twentieth week, they will either later not kill, or make a bad job of it. However, just seeing a prey animal is insufficient stimulus for the cat to kill the animal, the extra stimulus at this stage seems to come from the rough and tumble competition of its brothers, sisters and even the mother, who will either take away the prey animal or pretend to, to goad the kitten on.

Many species of cat play with prey they have killed, but Leyhausen feels domestic cats playing with live prey may be retaining their juvenile behaviour.

Does this ring a bell? Who hasn't played a similar role with a cat or kitten and a piece of string, playing the part of a 'competing' mother. However in situations like this an adult cat is far from being an unwilling stooge, for cats can take great delight in games, as they only too frequently show when having caught a mouse, they carry it unharmed to a position a few feet away from any cover for the mouse and then let it go, only to pounce just before it escapes (although not always!). A few months after our house-cat Jeremy moved in with us and she was still establishing her territory, she was engrossed in such a game when the cat that she had usurped who had been thinking of moving in, dashed in, grabbed her mouse, and ran off, leaving a bewildered and grumpy Jeremy. It was some considerable time after that before she would consider playing with her prey, and ate anything she caught rapidly.

The young kitten in the normal order of things sees and eats dead prey brought by the mother before he has to catch his own prey. If this is not the order of things, as in many domestically reared cats where the first food is 'caught as a tin', then on killing a mouse that is covered with fur, the cat may not associate the dead prey with food until it is 'opened'.

Feral cats and wildlife

Feral cats are part of our wildlife even if largely unrecognised. They are significant predators in rural areas and scavengers in urban areas. In many ways their role is similar to that of the fox, with whom they must be considered competitors, controlling each other's population by eating into the available food supply. Although appreciation of their role may be new, they have been exerting their influence on the rest of our wildlife even longer than the rabbit has been in Britain, and so are hardly new predatory competitors.

Feral cats and red squirrels

In Britain there has long been an emotive issue over the supplanting of red squirrels by introduced grey squirrels. It now seems that the red squirrel is holding its own in coniferous areas (such as Forestry Commission plantations in East Anglia, the Lake District and Scotland) while in deciduous hardwood areas the grey squirrel has ousted the red. In countries where both types have long existed together this pattern also occurs as the red squirrel is really better suited to softwoods.

From a study carried out in an area of deciduous woodland and fellside in Cumbria over a ten year period it seems that feral cats may have been keeping a check on red squirrel numbers. Feral cats were trapped near an isolated house and removed over this period and the population gradually fell. Red squirrels that, although previously living in the area, had not been seen for years, began to be seen again part way through the feral cat trapping programme and to increase so much so that in one year (1976–7) five dead red squirrels were found on one mile of road below the woods killed by vehicles.

Rats and mice – work and play

Farmers wanting rat-catching ability out of the farm cats have historically believed that keeping them a bit lean and undernourished will make them sharper at catching rats.

Farm cats have often lived wretched lives. The only feeding received by such cats has been milk given at the cowhouse during milking. However, if farm cats are well-fed cats they will be more dependable as effective ratters. This does not mean that ill-fed farm cats will not support themselves, far from it, but they are more likely to stray, having less attachment to the buildings and will hunt farther away in fields and woodland. (From the recent findings of home ranges and amounts of auxiliary feeding, it is clear that cats will indeed stay, more productively, closer around the farm buildings with extra food.). The effect of rats on farms today is not so apparently devastating as when unthreshed corn was ricked for late selling. Nonetheless, farm losses through damage by rats still runs into millions of pounds and has the associate health risk to farmworkers of leptospirosis.

Although it has been found that cats are ineffective on introducing them to heavily overrun premises, if the rat numbers are initially brought down by other means, the farm cats are then able to keep the numbers down.

Many farmers consider certain types of toms as 'prize ratters' which they may well be, but considering the larger home ranges of toms over queens, they may not be such effective ratters for the area that should be kept without rats, that is the farm buildings area. Research by Macdonald and Apps showed a larger hunting range for their tom relative to their queens which largely stayed near the farm. Perhaps economically farmers should keep a higher proportion of queens than toms. In a recent survey of 211 Yorkshire and Humberside farms, 198 farms had cats and in 88 per cent of these the farmers thought their cats played a worthwhile anti-rodent role, so much so that over 20 per cent of these did not use any other controls. However in complete opposition 10 per cent of the farmers felt cats had little or no use as rodent operatives. The 'What the Cat Brought In' survey however supports the pro-cat farmers, for farm cats were shown to be particularly adept with young rats.

Rats and cats

The relationship of cats and rats is a curious one. Cats recognise rats as prey while at the same time recognising their formidable nature. Adult rats can easily attain the size and weight of a weaned kitten which only has the advantage of longer legs over the rat. It is not surprising therefore that many adult cats are very wary of rats.

There exists a strong suggestion that the cat's ability not only to catch rats, but also to eat them, depends upon experience gained from its mother while in the confines of the cat's nest. To the cat the coarse hair of the rat (unlike the softer, shorter and more easily penetrated coat of the mouse) is not automatically associated with the meat of the rat. It is only perhaps the action of the mother cat in initially opening the hair-packaged carcase that forms such an association for the young cat. If this is correct those cats starting life without a mum enterprising enough to bring home the fatted rat would seem to have a decided handicap.

Having noted the incredulous stares of several cats on being presented with a complete dead rat, I ran a feeding trial to gain some insight into the relationship of rat meat to rat skin and hair.

Four groups of largely displaced domestic cats were presented with small cubes of fresh rat meat and separately with small squares of fresh rat skin (the skin was totally devoid of any meat and was presented with the hair facing the cats). Seventy-four cats were involved in this trial that took place over two days separated by a two-day interval. On the first day three of the groups took part in the trial just before their normal feeding time, while acting as a check the fourth group took part in the trial after eating. On the second day the reverse took place, with the three groups being fed before the trial, with the same fourth group this time taking part before being fed. The animals were not restrained in any way but were free to move about in their large group pens that they had lived in for some time before the trial. Indeed eight animals (in addition to the seventy-four) refused to take part out of shyness to the operator and were left alone.

A range of interests in these food items was clearly seen. A known experienced 'prize ratter' would eat both the hair and the meat without a moment's pause. Some of the animals that refused the offered samples, nonetheless sniffed it, then tapped it gingerly with a paw and licked their lips in a semi-nervous, semi-anticipatory way and while not eating it nonetheless continued to regard it carefully. A few, after following the above sequence and not eating it themselves, nonetheless would stand over it and growl threateningly at any other cat that came near. In contrast a number after a cursory sniff had no further interest. Those who did eat generally showed a considerable interest in the samples, some excessively so. No cat played with a sample. It was clear that a large number of cats were prepared to eat rat as food (the range was from 30–86 per cent eating depending on circumstances). It also was clear that fewer cats were interested in eating rat meat after previously eating (a proprietary cat food which was not restricted in amount). This difference was particularly noticeable in the responses to meat as opposed to hair-covered skin. For three of the groups although a fair number ate the skin, a larger number would eat the meat if they had not previously eaten. If the group had already eaten then the percentage of cats eating rat meat dropped to a similar level to those eating rat skin.

The interest in rat meat seems to be inhibited by previous eating, but the interest in skin, although less, does not seem to be affected (except in animals unable to use their sense of smell). It is perhaps possible that the unvarying skin feeders were those who were initiated into the food containing significance of rat skin and its characteristic smell by mother.

Some balance to the initiation theory of a cat being introduced to a rat while a kitten may be met by the experience I had with a nine week old black and white kitten in a different trial to the above. This kitten had been born from a very wild feral cat, captive in a pen, and there had been no opportunity for a rat to have been presented to the kitten. Nonetheless when offered both meat and skin it readily ate them. I developed a soft spot for this kitten for it was a bold loner, totally fearless, as I discovered when it leapt out and tackled one of my shoes (with me in it) by flinging both front paws around from the back and sinking in its needle sharp claws. While by no means disproving the initiation idea, this kitten clearly demonstrated that at the very least it is not obligatory.

Some cats become particularly expert in ratting and the honour of the highest recorded tally belongs to a female tabby who worked the White City Stadium in London from 1927–33 catching 12,480 rats (that is an amazing daily average of between five and six per day over six years).

Cats catching rats

The cat uses a special technique when confronted by a live rat. I have seen the natural encounter between urban cat and rat, within a couple of feet of where I was standing. The rat ran along the base of a wall and found its path blocked by the cat. The rat lifted its head and opened its mouth wide, threatening the cat with its teeth, shaking its head slightly side to side. It was a large brown rat and the cat was only of moderate size, so this was a serious threat and the cat took it as such, backing off. The teeth were menacing almost level with the cat's chin. The cat spun on its haunches to be alongside the rat's head, which it then proceeded to beat a tattoo upon with its front paws. It was now the rat's turn to spin around and it tried to go back rapidly the way it had come, giving voice to squeaks as it went. The cat meanwhile beat until the rat had moved from the range of its paws, then coming down onto four feet it took a few paces to come alongside the rat and beat it with its paws again. It did this until the rat had almost reached its escape home, when the rat, perhaps fearing this escape was being cut off by the cat returned to the original route, pursued with beatings by the cat. The change of route along the wall was repeated once more by the rat, but it was not able to shake the cat.

Compared to the quick kill of a mouse or bird this seems at first sight far more elaborate and specialised. However, although the killing bite to a mouse is quick, depending on factors already outlined the pre-death and post-death play consists of types of actions that are required to make the rat sufficiently submissive that it can be killed. When applied to a mouse the smallness of the animal makes the movements seem exaggerated and unnecessary. The cat will often sit on its haunches and, with arm extended, tentatively pat the mouse a few times with claws extended. With the extended raking claws, when the cat lifts its arm back to sit down, the result can be that the mouse was hooked and is thrown spiralling through the air. Having used a few rat-controlling pats on a mouse the latter will usually sit there stunned, so the cat will stop and wait to see if there is any movement before proceeding, and normally looks around. What could seem like unnecessary cruel play can be seen as part of the normal hunting gamut onto an animal that may hurt the cat and need subduing, but is naturally not seen in its proper context, as fewer people have watched a rat being chased by a cat than those seeing a mouse followed.

Such movements of small prey have the advantage to the cat of practising its skill for when it is desperately needed, for after all the size difference of a large rat to a cat is not too dissimilar to an Alsatian or wolf and us. There is an underlying automatic response, partly learnt, but automatic nonetheless to a prey animal and it is a catching and killing method that is triggered and still includes the patting but to a far lesser extent for mice as it is obviously less required. Nonetheless were I in a cat's position, I am sure I would bat a mouse a few times before going in with my face to bite it, as even a small cut in the wild leading to infection can be very serious. Fair play could be fatal! The patting ensures a dazed prey with its head low enough for the bite to be administered. Further, as anyone is well aware who has ever watched a cat approaching milk in a saucer, while their field of vision is fine at a little distance (to find the saucer) it is hopeless at things close to under its nose (finding the milk). A cat moves its head cautiously down towards the milk, and is only stopped by the sense of touch when the lips meet the milk and usually stops with a small but perceptible judder. Bearing this in mind it makes it more important for the prey to be dazed.

Play, in the true sense, probably comes into action when the cat is confident and hunger and other factors are minimal. The hunt and dazing movements can then be skilfully used, for example in facing a mouse. Nonetheless, the movements are those of dazing prey, even those movements where the mouse is batted back and forth a few times in quick succession between outstretched paws, with cat on haunches. These are the same actions as beating on a rat, seemingly changed, but only because it is against a smaller prey. This set of double-side batting is often encouraged by people living with cats, by providing ping-pong balls, marbles or something similar.

An economy of movement exists between the hunt and the dazing actions. The head held back and single paw batting is very similar to exploratory probes into likely holes and crevices when hunting and will become more vigorous leading directly into batting and dazing if a mouse is found. This can also be seen when a cat is playing with a mouse, when the cat will place itself intentionally on the other side of an object from an already caught mouse, so that an exploratory paw can be put around the object, such as a chair leg, to bat the animal from that position.

Cat and mouse

One thing that even ardent cat admirers find hard to reconcile is the domestic house-cat's habit of playing with its prey, when it is armed with the means of rapidly and efficiently dispatching the victim.

Leyhausen has suggested that the cat, like most animals, is subject to certain drives and that for survival some rank higher than others. He showed by releasing one mouse after another into a cage with a cat that the cat was always ready to catch a mouse even when it had eaten so many it could not eat more. The cat would catch the mice and with a number dangling from the mouth and one under each forepaw it would still want to catch a new one. The urge to catch was still keen even when it had no urge to eat. It is as well the urges are in that order, for if they were not the cat would starve. It is the same balance that makes the house-cat run to the fridge whenever the door opens, even if it is not feeling hungry. It still feels the need to have enough food available.

Farmers say they can not understand why a fox slaughters all the hens in the henhouse at one go – but it is only the same need in the face of an abundance. People similarly buy more food than they need for the next meal, as any larder shows.

It has been suggested that wild animals desperate to survive bolt their food, not wasting time playing or chewing unnecessarily, so they are ready to look for more. On this basis rural feral cats seem to act as wild animals, for in one large feral tom I found three small mammals just bitten through at the waist and swallowed. Despite this Macdonald and Apps found their female Devon farm cats played with prey.

I did not release live mice, but I did give dead mice to see how the manipulative drive or play related to eating. I found a cat would eat and eat mice to the extent that I had to stop giving the cat any more on each test day, even though it sought more, for fear of rupturing its stomach. Before eating the first mouse the cat would have a period of play, feeling the mouse with its teeth. After receiving and eating the first mouse the cat would undergo vigorous long-tongued washing and extensively washing its jaws with both paws. But by the time the second mouse was supplied immediately after the first one was eaten, the play time had fallen considerably and the post-eating wash was less vigorous. By the time the third mouse was supplied the washing was just perfunctory.

Unlike the bisecting action of the feral cats, domestic house-cats sit down to a leisurely and systematic meal. The head is always eaten first by taking it in the mouth and shearing it off at the neck. It is usually chewed a bit then swallowed as a lump. Next the mouse is sawn in two by the teeth at the level of the diaphragm. The top of the mouse (including lungs) is then slightly chewed and swallowed. The cat will often pause and look around between each section. It then tackles the lower half of the body, most usually starting from a back leg, eating back to the intestines, often discreetly leaving the colon.

The more mice the cat consumes in any one day the less time it takes to eat any bit (until after the fifth mouse it begins to tire).

Unlike the eating, the pre-eating play period falls with recent experience. If a cat has not eaten mice for a while the first mouse play period is longer the first day than the second. This would have value in the wild, for if there were a glut of mice, the less time spent playing the better. Regular feral mouse-hunting cats would play less with mice than the occasional household mousers. Any element of sharpening skill in order to keep the eye and paw in means the more you are catching the less you need to practise.

The question is not just 'why play?' but why more often than not with female cats, 'why take it back to the core area and play?'

It cannot just be for security. Whether it is taken to a feral/farm core of other cats, or into the domestic home, the risks of it being removed from the cat are higher in either place than eating it on the spot.

With human 'conspecifics' and even some other cats there may be a maternal element involved. The home-cat knows we are cats, but we seem a bit backward in catching our own prey, so she will mother us and demonstrate the hunting skills as she would to her own kittens.

For whatever reason a cat plays, even if it is just for the sheer joy of the chase and 'recapture', it will serve a useful group cohesive function. By group members effectively displaying their prey, others are shown that this is a good place to stay.

Bats and cats

The cat as supreme hunter is probably shown most clearly with bats. The superb flying sonar-equipped bats have few predators in Britain. From a study of greater horseshoe bats there is no record of bats being eaten by owls, despite the hunting efficiency of owls. Nonetheless most recorded predator deaths are by domestic cats, when the cat has presented its owner with its catch. Bats do contract rabies and in 1980 in continental Europe, and earlier in the USA, there are records of cats catching rabid bats and taking them home to non-rabies areas.

In Somerset a family of feral cats lived in a cave and preyed on the bats as they flew through a narrow entrance. The bats were only part of the cats' diet, but as the bats were ringed, these events were well recorded.

Shrews and cats

The shrew seems to fit an unusual role as a prey species for the cat. This results however from the apparently unusual characteristics of the animal. Shrews are regularly caught by cats but are rarely eaten by them. On examination of what had been eaten by cats in Wisconsin, North America, it was found that shrews only occurred in 2 per cent of the cats. When they are found in a cat they usually occur singly. However one cat did sufficiently overcome such apparently universal distaste for shrew meat to be commemorated in the scientific literature as actually 'making a special effort' to seek out and eat shrews. This large Illinois tom was found to have eight short-tailed shrews in its stomach when examined.

An explanation for the low number normally eaten is found in the shrew being a voracious carnivore. In classification terms the shrew, although carnivorous, because he feeds on invertebrates is termed an insectivore (along with moles and hedgehogs). Due to their high protein diet carnivorous-feeders' flesh is generally unpalatable to fellow carnivores. In consequence it is not just cats that turn up their noses at the thought of a shrew, indeed it may be that the cat is more tolerant of shrew in its diet than many other carnivores. An examination of the diet of racoons, red and grey foxes, skunks, mink, weasels and opossums as well as cats, found that more shrews were eaten by the cats than by any other predator.

The ease with which cats catch shrews also relates to the shrews highly carnivorous nature. To obtain sufficient food the shrew hunts during the day and the night with a series of frequent activity peaks. This increases its exposure to prowling cats over the mainly nocturnal wood or field mice and the mainly day-light-active bank voles. The shrews urgent search for small prey means that it hunts often oblivious of being hunted itself to the extent of quite noisily bustling around.

Their movement and rustling of leaf litter make them a major food prey item for owls and although cats do not eat many shrews they catch and kill a great number, sufficient to be considered major predators on these animals. This tendency has been usually noticed anywhere that shrews and cats meet.

The time of the year has relevance, for many older adult shrews die in the autumn and are fairly easy to find dead on rural paths. So much so that during the making of a film of wildlife in an east London churchyard, I lost my balance rather than stand on a dead shrew on the path, and stumbling, accidentally put my foot (and leg up to my knee) into a grave! It was a very old churchyard and very overgrown, and I had not seen that the stone cover of the grave was slightly ajar. As it was dark and having experienced the proverbial feeling of having 'one foot in the grave' I left rapidly! However I had noted the dead autumnal shrews and prowling cats from nearby houses.

It has been suggested that the older adult shrews are less capable of holding a territory against the new generation and so die off in the autumn.

In Britain I have found my house-cats find shrews a particularly easy prey. When I lived in Somerset our brindled tortoise-shell neutered female, Sukie, caught more shrews than any other small mammals (being about 80 per cent by number over a three year period).

This occurs to such an extent that I feel captures (but not consumption) by domestic cats are probably as good indicators of shrew density as any other method (certainly doubts exist as to the measures of density given by live-trappings methods with shrews).

A number of other prey items have been found to be eaten as often as the insectivorous shrew and yet are generally not considered to be distasteful, but rather desirable prey items. This apparent paradox is rapidly resolved by reflecting that these latter items are eaten at the frequency they are caught, while the shrew is often more numerous, and certainly more easily and often caught, but nonetheless eaten infrequently.

An American naturalist living on a farm in Michigan noticed that although his cat would take other prey in, the cat obviously classified shrews as a different type of prey for these alone he deposited uneaten on the porch.

Sukie operated a similar classification system against shrews in Somerset. Our garden merged into both scrubland and woodland, and she regularly brought prey home from her forays, indeed probably most of her catches. The catches she seemed proud of, or we could rationalise that she considered edible for her 'wayward kittens' (i.e. us) were either brought into the house or put in her 'mortuary'. Shrews on the other hand, of which at the height of the season she would catch five a day, but more normally one or two, she left in a perfunctory manner littered outside one of the three external doors of the house. These she would either leave intact or more usually just chew their heads off.

The 'mortuary' which she maintained I found fascinating. It was underneath an evergreen tree and was about twenty feet away from the house. The branches of the tree came down to the ground keeping a diameter of about six feet of soil totally dry and hidden from sight. The corpses that she laid out in this safe place were intact and amazingly laid side by side on the soil in a line, with about half an inch between them. Of a summer's day one of her favourite snoozing spots was beside this tree. When we moved house in Somerset (although only a half-mile) the new place only backed gardens instead of woodland and scrub, and the shrew catch plummeted to zero.

Our current house cat Jeremy lived most of her life in East London where, whenever prey was available, she would catch it. She would systematically catch fledglings and young mice emerging fresh from their nests, which were then brought near to us. Never a shrew. On being taken some thirty miles into central Essex to live for a while in a house with an ungrazed field she proceeded to catch three shrews in the space of as many hours on her first afternoon in the field. This certainly reflected the relative availability of shrews in the different habitats.

Cats and moles

Moles share with shrews the insectivore order and the same fate of being caught by cats, but rarely eaten. Indeed it seems that in Britain the two major predators of the mole (excluding man) are the fox and the cat. Moles figured among the animals laid out in Sukie's mortuary, seemingly kept more out of pride than for any edible quality, and in keeping with this moles were never taken into the house. On one occasion I witnessed her detect and catch one of these in its tunnel. In a lawn disfigured by an array of molehills, she became interested in one spot. Within a relatively short space of time the surface of the ground began to move and a molehill began to appear. After watching intently for a few minutes Sukie began to dig into the appearing molehill. She then pulled the mole out head first.

This is of course the correct technique to employ to remove a mole as you instantly discover if you try to take a mole out, as I have, the other way round. I tried to pick one up, by the rear end, as it was just beginning to break the surface of the soil. I saw that the only way I could have lifted him up from that direction would have been by pulling to the extent of damaging him, so strong was his grip with his front claws that continued to steamhammer into the ground. Needless to say I let him go. The cat's method of grabbing by the nape of the neck, extensively used against other small mammals, remains successful here for the same reason that a fishhook is only removed in one direction. Although she caught moles, Sukie would not eat them.

This repeated catching of unpalatable insectivores by cats would seem to give further backing to the idea of drive ranking. Hunger has been shown to be satiated long before hunting. In this case cats continue to hunt prey (insectivores) they would not wish to eat. Even if they don't eat moles, cats' ability to catch them and keep gardens 'mole-free' was recognised and recommended as long ago as the fourth century AD by Palladius.

Rabbiting

A cat does not have to revert to the wild to catch rabbits. Sukie, who for most of her life alternated between sleeping and hunting, is I am sure not the only domestic cat to make a habit of bringing home rabbits both larger and heavier than herself. The back of my home in Somerset faced a rising garden which petered out into a hillside of gorse and bramble, which in turn merged into woodland. The hillside was ramified with the tunnellings of rabbits, with entrances both in the gorse thickets and in the cropped-grass stretches between. I was able to watch the expert hunter on these exposed areas. She would on occasion flatten herself beside the entrance and incredibly fasten onto the neck of emerging rabbits. Perhaps equally incredibly the rabbit would go limp. Sukie would stagger down the long walk to home with a marching stride, tensing her own neck muscles, keeping her head erect and slightly back, so that her huge burden would not drag on the ground. On arrival at the lawn by the house, she would present us with – a live unharmed rabbit!

After tickling Sukie and praising her, we would pick up the rabbit and return it the long walk back to its home. Unfortunately it did not occur to me at the time to mark the rabbit to see if it was the same one returning each time, or whether they 'took it in turns'. I must admit to reading a resigned look on the rabbit's face of 'here we go again'!

It is believed that the rabbit population from the time of its introduction into Britain (during Norman times, later than the domestic cat) until the early nineteenth century remained at a low level kept in check by natural predators. The introduction of the Ground Game Act of 1880 giving tenants the right to the rabbits on their farms, and commercial production of the rabbit gin-trap, in combination, led, it is suggested, to a rapid increase in the rabbit population. The traps used to catch some 10–15 per cent of other animals, largely predators, including the cat. Commercial interest in maintaining the rabbit and reducing the predator triggered this increase in rabbit numbers which continued up until the advent of myxomatosis and the banning of such inhumane traps in the 1950s.

An agricultural zoologist writing in 1946 saw farm cats going rabbiting, where often by the actions of the rabbit catcher or keeper they were killed or maimed.

During the time of some experimental infestations with myxomatosis carrying rabbit fleas, an island community of monks, driven to distraction by a huge number of rabbits reducing their agricultural effectiveness, attempted to check their numbers by releasing a large number of cats. However the monks found that the cats also killed their pheasants. When myxomatosis was introduced the rabbit population slumped. The monks then had to resort to shooting the now starving cats and by so doing demonstrating that the cats had previously been living on something more than pheasant.

Cats stalk and leap upon rabbits, taking them by surprise, especially the young when leaving the nest stops. The majority of domestic cats are not powerful enough to catch hold and kill a healthy full-grown rabbit, although they will drag home a weak, diseased one.

Healthy adult rabbits have slashing teeth and powerful kicking feet with sharp claws. Nursing does have been seen to drive large cats away from nesting stops containing young.

Staying on a farm in Huntingdon at a time of many young rabbits in surrounding fields, I noticed that this particular farm's cats were not just sleek, they were positively fat from living well off this rabbit surplus. They would cull the young rabbits (disconcertingly called 'kittens') in the fields and bring them back to favourite lying spots within a forty foot radius of the farmhouse where they proceeded to crunch away audibly.

Mr Hambly-Clark an Australian hunter in Adelaide said in 1971 that cats unlike foxes, even went down rabbit burrows and killed rabbits. However, from Australia's history with the introduced rabbit this action is unlikely to stir much reaction.

Nonetheless the South Australia Conservation Minister understandably pushed for investigations into the action of the cats on the native birds in the national parks.

Towards the end of the last century an attempt was made to introduce a colony of rabbits into London's Hyde Park but urban cats moved in, and eradicated the rabbits.

Cats and birds

To judge from their expressions when a stalked bird flies away, cats never seem to have fully understood flying. They are capable fledgling catchers, but when birds are competent at flying, cats are less capable at catching. However, each time a very skilful stalk is employed and on rare occasions, effort is rewarded. Most songbirds, by having eyes at, or towards, the sides of their heads, have a wide field of view and in some a full 360 degrees. This enables them to see even a cat stalking up behind and so more than stealth is required by the hunter. The cat is well aware of this and so its method of serious stalking is for the body to sink down between the limbs such that elbows vie with the scapula blades as to which is highest. The head is kept low. Although slow stalking may be employed, success usually accompanies a dash and freeze variant of the party game, with a catch as the prize. The cat stays frozen until some object, due to the bird's activities, blocks the bird's view in the cat's direction. Then the cat sprints to a point where the bird may be able to see it, and freezes again – or if close enough risks all by rushing forward. Although the numbers of birds caught by cats are small (allowing for the exceptional specialist cat) the numbers eaten particularly by domestic cats are smaller. As with the caught rodent's fur the immobile feathers can inhibit eating, especially with the larger species of bird with stiffer feathers.

From the cat's point of view not only do birds not play fair by flying and having eyes that can see beyond the back of their heads, but they can positively cheat by using loud alarm calls and throw the cat's chances of catching any others. The European blackbird is an arch-demon in this regard. Contrary to popular belief few bird species sing throughout the year in England but the peak of activity coincides with nesting and rearing young. When a blackbird realises that its nest-site has been detected by a cat it defiantly chinks out its repetitive alarm call. This is often sufficient to deter a cat. If it does not, it will reinforce its threat by diving a number of times sufficiently close to the cat to part the hair on its head! This latter is more than just a figure of speech on my part, for I have witnessed this when a cat had dared to approach within six feet on the ground of a blackbird nest that was only five and a half feet off the ground.

Although in the context of large avian populations the predatory cat does not take a heavy toll of the birds, nonetheless as an introduced predator (Chapter 11) on a restricted species, for example on some islands, the same is not always true.

Many people fear the devastation of their bird population rather than their rodents. They may have a higher responsibility to the bird population than they think, but not immediately due to the cat. Most of the birdlife of the British Isles evolved to fit our landscape when in effect it was one huge woodland. The piecemeal removal of trees for agricultural development over the centuries was balanced by the establishment of hedges in fits and starts. Hedges act as very thin woodland maintaining an artificially high level of birdlife in an otherwise biological desert. However below a certain length of hedge most birds will not tolerate the conditions. The countryside at dawn is becoming quieter in its chorus, especially in eastern England and other areas where huge lengths of hedge have been ripped out. Today the dawn chorus is often likely to be noisier in the bushes and trees of gardens in towns and villages than among the fields. Many farmers may have shirked the responsibility, but it does mean that it is off-loaded onto the gardener's shoulders.

This all being so, does the domestic cat and its strayed colleagues decimate the bird population in our towns? Many think they do but the evidence does not really support this.

Our neutered female house cat, Jeremy, can certainly catch small mammals but like most cats is not so hot on birds.

For as long as people have kept cats it would seem that they have appreciated that cats can catch birds. It is often suggested that cats take a heavy toll of bird-life but is this based on firm ground? Our house cat only caught fledglings when they were beginning to try their wings. Then she would usually bring in a few fledglings each day for a brief period of about a week. There were usually only one or two nests available within her range each year. Nonetheless fledglings, as with most young animals, are extremely vulnerable, and most would succumb to one predator or another and populations are stable with an apparent colossal culling.

The British 'What the cat Brought In' survey showed a bulge of fledglings caught in early summer, and a larger bulge of small mammals in late summer and autumn. In rural Pennsylvania while the volume of mammals increased during autumn and winter, in spring and summer the volume of caught birds increased. But not all of these birds are fledglings, for a recent study of birds' ringed (under the national scheme in Britain) has shown that nesting and rearing time is also the most vulnerable time for adults from cats, for busy on the ground seeking food for chicks, they are more prone to capture.

The frustrated chatter of cats jaws in the mechanical staccato fashion at birds out of range seems to be an exercise in wishful thinking. At the point of a kill it is believed that the cat has fine control over its neck grip and due to generous nerve supply it has touch-sensitive canine teeth, allowing location of prey neck vertebrae by rapid fine chatter to the point of dislocation or crush. The tetanic reaction is fired by the desire to strike at the sight of the elusive prey bird sitting the other side of a glass window, or up out of reach on a roof gutter. This is despite its pre-emptive nature and any warning to the would-be prey, as the frustrated realisation grips the cat that it will not reach the subject. The chatter is an eloquent 'If only. . . !'

The significance of a triggering neck is reinforced by birds like owls that are unlikely to figure frequently in a cat's catch. Not only because of their size, but also because of the apparent absence of neck due to full feathering, owls seem to be of little interest to cats. Tawny owl fledglings are individually dumped by their parents as soon as they can fly one hundred or so yards from the nest, where they patiently sit all day waiting to be fed at night. My cat has sat nose to beak with such a fledgling and showed no more response than curious interest. No neck – no bite!

In both Australasia and the USA the eating of birds (including domestic fowl as well as songbirds) has not been seen to pass 20 per cent of the volume eaten in any population studied. This is so even in the truly feral populations of Australia, as with the cats in the outback in Victoria, where despite the huge amount of mammals being caught and eaten, the birds taken are only some 4 per cent of the total, a small percentage compared to other cat populations. In Oklahoma it has been found over three years that country cats compared to town cats caught more mammals but were much less of a menace to birdlife.

In Colin Howe's Yorkshire 'What the Cat Brought In' survey only a quarter of five hundred prey items were birds. Although rural and suburban cats took mainly small mammals, urban cats probably with less choice took mainly house sparrows. The top five birds caught overall in the survey were (in order): house sparrow, starling, blackbird, greenfinch and dunnock.

However, a caught bird is not only not necessarily eaten, but may still survive. Tentative 'play' patting by the cat tests if the prey is dead. A macabre game is enacted by *both* players. Although the original blows may have been to ensure submission, to enable the safe approach of a disabling bite, the cat will not begin to eat until the prey is dead. As its near-muzzle vision is non-existent, the cat would be anxious not to start leisurely eating, if that risked its lunch turning round and biting back. Small birds can 'play dead' ('akinesis') to fool the cat. Both small birds and small mammals employ this defence, for continual fluttering or running invokes the chase. So patting is not enough, and the apparent breaks of disinterest when the cat looks away may well be a counter-measure to see if the prey is shamming and may make a break for it.

Butterflies, earthworms and others

Partly digested insect bodies have been found in the stomach contents of Australian and American feral cats. In an urban area of Oklahoma, Orthoptera (grasshoppers and crickets) have been frequently eaten, while in a wooded area of Maryland, USA, it has been found that a fair part of trapped cats' autumn diet has been insects, again mainly orthoptera. In the rich agricultural lands of the Sacramento Valley, insects form about 7.4 per cent of the feral cats' total food, and are an important item in their summer diet.

Almost all of these were crickets (although a few other insects did turn up such as an occasional dragonfly). On the arid equatorial Galapagos Islands the adult feral cats are the smallest so far measured in the world. They have both a significantly smaller body weight in both male and female, and body size. This is due to both their diet and a lack of freshwater. Cats with their high protein diet need freshwater, or suffer kidney failure (as seen when domestic cats live primarily on dried foods). The staple diet of the island's cats is a combination of small larva lizards and grasshoppers. The cats gain moisture from the grasshoppers but relatively little protein. Scales from butterfly wings have turned up in the English feral cats, and among twenty-three samples of feral cat droppings from Portsmouth Dockyard, Jane Dards found about a sixth had eaten insects. Macdonald and Apps' farm cats in Devon were seen to eat fifteen species of vertebrate prey, and among many insects, again to particularly catch grasshoppers. Certainly many domestic cats will bound, hesitate, bound, hesitate, and bound after grasshoppers time after time, but are not always successful.

Our cat Jeremy will stand on her back legs and catch butterflies in mid-air then sit back down, scrunching away with relish, the insect exocuticle sounding like crisps being eaten.

'I saw him catch and eat the first butterfly of the season, and trust that the germ of courage, thus manifested may develop with age into efficient mousing'.

THOMAS HUXLEY (19th century)

Large shimmering bodied dragonflies act as immediate bait to Jeremy and certainly all other cats I have seen that are near one. They drop everything and pursue. Cats will also readily catch spiders and craneflies.

'. . . her watchful eye is most strange to see with what pace and soft steps she taketh birds and flies'.

EDWARD TOPSELL (17th century)

Studies of cats' stomach contents have on occasion revealed earthworms. However by the same source as above I know cats are capable of catching live earthworms. There is only one earthworm in this country that, when the soil is damp enough to allow it, will normally graze at night on dead leaves and twigs on the surface, and this is Lumbricis terrestris, the so-called common earthworm. They will always keep their tails down holes with their seta (claw-like bristles) firmly clamped into the soil there, ready at the slightest feel of danger to retract rapidly down their burrows. It requires great skill to catch them, but as these worms can be as long as a cat's tail they are worth examining if you are a cat. I have stayed out at night and watched Jeremy catch them by sitting in front of the head of the browsing worm and a little to one side. Then by a featherlight touch she hooked a claw under the worm's dorsal skin. The worm instantly retracted, but because she had hooked it from the front it could not escape because of the curve of the claw. I have since come across a number of other house-cats that catch worms. Cats have the capability of easily catching earthworms and in feral cats, certainly in some parts of the world, they contribute to the diet as the cats and worms will both be out looking for food at the same time.

Reptiles and amphibians

Cats will readily catch snakes when about, and my brindled house cat Sukie in Somerset would on occasion bring in a grass snake or slow worm. The odd grass snake also infrequently occurred among the Yorkshire 'What the Cat Brought In' survey of house cats' prey. David Macdonald and Peter Apps' Devon farm cats were also seen to kill a slow worm (legless lizard). Mammals were the major prey item of 219 free-living cats of the Sacramento Valley, California, of which they caught nearly four hundred, in comparison the reptile catch amounted to thirteen snakes. To the feral cats of the Galapagos Islands lizards are more of a staple feature of their diet, for the island's lava lizards are of a prey mass similar to a mouse, and are virtually defenceless.

The hopping movement of a frog will catch the interest of a cat, but its angular stocky shape and lack of obvious neck position can cause a similar halt to the proceedings as the apparent absence of neck of a tawny owl fledgling. Nonetheless certain cats overcome this problem as in the case of Kiki, a twelve year old neutered house tom who was something of a pond specialist. Fish sheltering under lily-leaves were caught by a fast rake with the claws that left

Right: Urban feral cats despite living close to man do not appreciate captivity. Cat hiding in small space between door and wall of pen

Below: The related Fitzroy Square female cats in brief captivity following neutering huddled together behind the large protective queen (in contrast to *below left* p. 6)

Above: The free-living feral cats of Fitzroy Square feeding behind the safety of the square's railings (see p. 90)

Left: It's not just on TV commercials that an enterprising cat can eat with its paws. One of the Fitzroy Square cats feeding

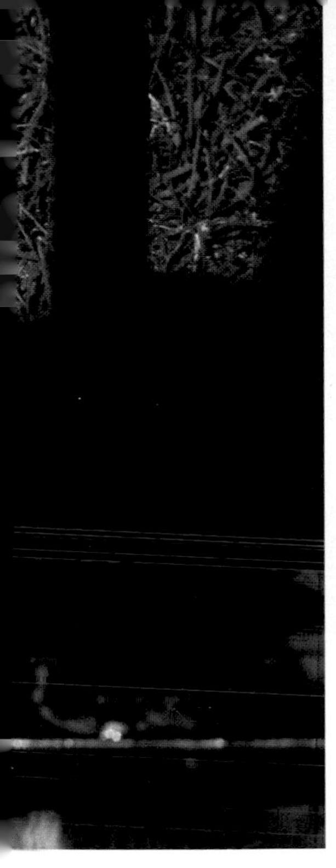

Above right: In the absence of trees, fences and even cardboard boxes may be marked. Urban feral cat's marked sunning box

Right: Tree scratching forms marks, stretches muscles, but it also sharpens claws. Curved slivers of keratin flaked from the sides of the claw remain in the bark (see p. 135)

Female house cat ("Jeremy") chinning on the ground (see p. 144)

Cats greeting prior to head rubbing. Cat seated is a blotched tabby (see p. 29)

Right: House cat with mouse in neck bite (see p. 114) *Inset:* House cat 'test-playing' with mouse (see p. 114)

Above: House cat with killed frog (see p. 128)

Left: Frog in absence of clear neck was killed at the throat

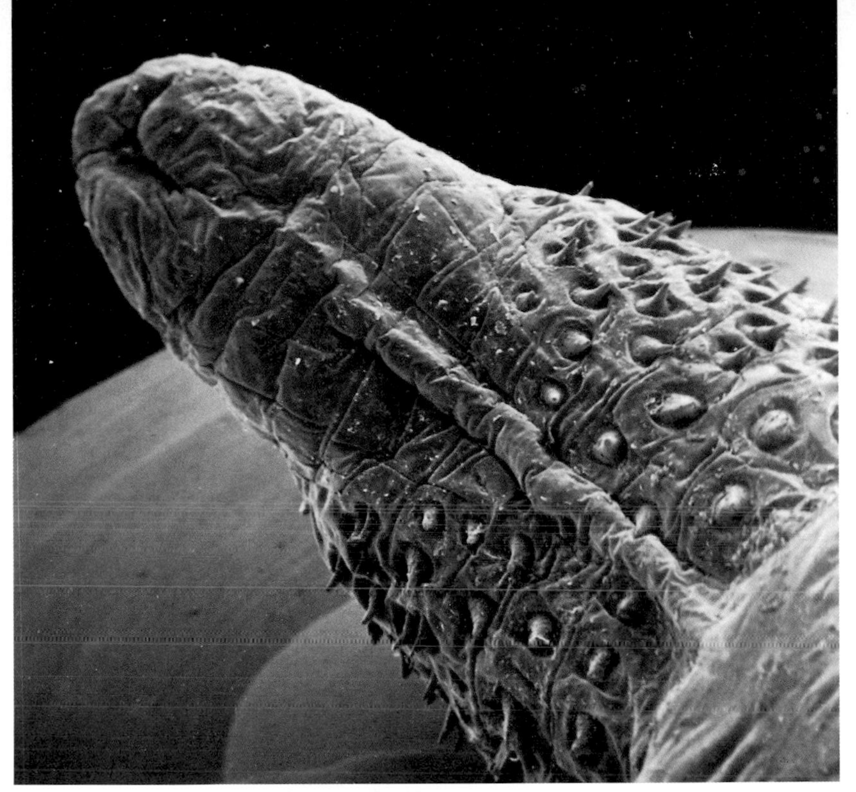

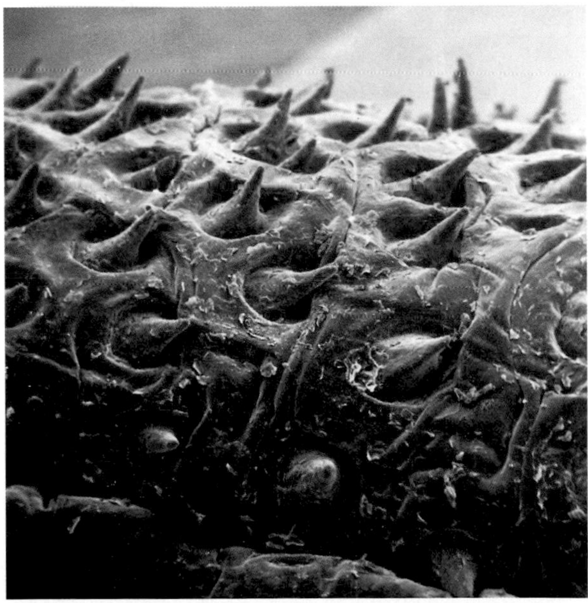

Above: Phallus of tom cat showing both its curved shape and barbs

Right: The barbs on the male's phallus that induce ovulation in the female cat. Their size depends on the hormonal state of the tom (see p. 171)

Left: (Above and below) The same 'scent stick' being investigated by two different cats (see p. 138)

Below: The rough surface of a cat's tongue seen by scanning electron microscope (see p. 35)

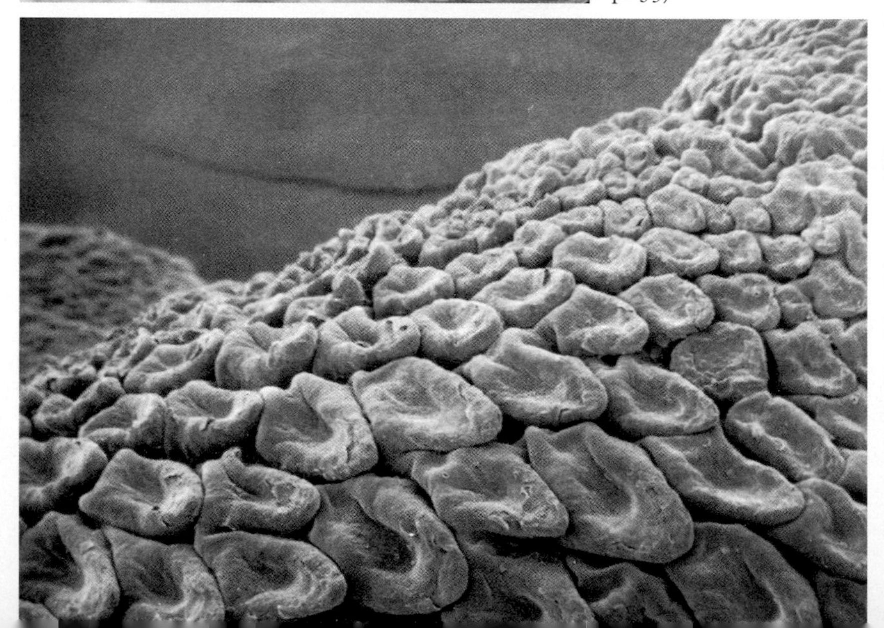

neat parallel lacerations through the floating leaves. Frogs have more of a visible neck on their underside, and although throughout the cat family only overlarge prey normally receive an asphyxiating throat-clamp rather than a neck bite, Kiki killed at the frog's throat.

Cats and fish

I have yet to come across a group of feral cats that are believed to be existing *entirely* on raw fish, although this may be small comfort to the depleted garden pond owner! However, I have come across one feral cat that had over 300g of smoked haddock crammed tightly in her stomach. It is just as well cats do not exist only on raw fish as cats that were experimentally fed in the 1960s just on raw carp and herring lost weight, showed erratic movement and died within 23–40 days. Raw fish contains an anti-thiamine, something that acts against the effect of thiamine (vitamin B_1). In people the condition of beriberi induced by lack of vitamin B_1 also causes death if untreated. A lack of thiamine profoundly affects carbohydrate metabolism.

The dustbin ecology

When I lived in Somerset I was fortunate to be able to observe our cat Sukie, the accepted woodland predator, returning home from hunting *through* a family of playing badgers on the lawn just behind the house. The badgers would come and raid our dustbins and very noisily strew the contents over a large area. For months we might not see them then for long stretches at a time they would come every evening on their forages, down a very definite path. Needless to say we tried to prevent the ransacking of the bins, by bricks on lids and roping down the lids, but to no avail. The persistent brute strength of the badgers won out. We resorted to bribery, trying to fill them up before the dustbin assault by putting dishes of food on their path – and after a fashion this was successful. It would be during the extended play period around the food bowls that Sukie would calmly stroll through the group, neither she nor the badgers giving any noticeably outward sign that the other was there despite being only feet apart.

For certain animals in England, especially in urban settings, a group of dustbins have become the equivalent overlapping focal point in movements to the African water-hole for big game. Dogs will scavenge by bins in daylight hours and this is not only noted in England but well documented in America by Alan Beck in a study in Baltimore of stray dogs. Unfortunate down and out people join in the sifting of garbage.

As daylight wanes the dustbin as an urban ecological focus really exerts its magnetic pull. Cats, badgers and foxes all overlap in their movements.

This nocturnal trio of scavengers and hunters is also working the same area as yet another in the same category – the hedgehog. At first sight competition may seem to be slight in that the hedgehog seems to be after smaller prey like worms. No so, for not only are earthworms the staple diet of badgers, they also figure in that of the fox – and the cat.

In the urban wilderness the free-living animals that flourish have very special qualities and are largely of two types. The first are the feral animals, once domestic and now wild. This includes not only the feral cat, but the feral pigeon as well. Although the distant ancestors of these birds were European rock doves, they had been domesticated and bred by pigeon fanciers. The occasional bird that made its bid for the wild, joined up with others and with a bit of breeding gave rise to great flocks.

Although the cats have a gastronomic interest in the pigeons the latter do not often allow themselves to get caught. When watching the urban feral cat elaborately stalk and then try to run pigeons down with short rushes of speed, in my experience the bird usually escapes! Nonetheless, I have found the 'give-away' signs in the form of groups of pigeon feathers stuck onto the bloody spot in the favourite resting place of members of a group of feral cats, in the ducting underneath one of south London's hospitals.

The second group are wild animals that have not suffered domestication and includes foxes, hedgehogs and to a lesser extent badgers. Foxes have now penetrated into the heart of the capital, nevertheless it is in suburbia that they are firmly established.

Surveys of hedgehogs in east London have shown the animals to be far more numerous than previously believed. The nocturnal ramblings of both cats and hedgehogs overlap so much that people trapping feral cats often find hedgehogs inside their traps unconcernedly eating the cat food placed on the treadle. On the Bedfordshire/London border one hedgehog kept getting trapped. It was so appreciative of the cat food that it kept returning and setting off the cages, to such an extent that the operative was forced to put it in a cardboard box for the rest of the night with some food and let it go the following morning when trapping was over.

The one feature that these two groups have in common that gives urban ecology its unique flavour is that although a healthy caution is shown of man, neither are overtly intimidated by man or his associated structures. Indeed these structures have survival value in the form of garden sheds standing on bricks and office 'temporary' buildings on foot high pillars, which provide shelter for cats, foxes and hedgehogs. Gardens with permanent pasture (lawns to us) furnish earthworms for hedgehogs (as well as for foxes and cats) and office blocks make splendid surrogate cliffs for ex-rockdoves.

It is sometimes suggested that because urban wildlife survives we need not be concerned by the reduction of wildlife in rural England caused largely by changing agricultural practice. What is overlooked in this argument is that as a smaller range of species is seen in towns (shyer animals such as fallow deer have probably abandoned even Epping Forest (by London) as 'people pressure' has grown) and the fewer species are biologically less stable. However, if towns are not thought of as a wildlife alternative to the country outside, but rather as a specialised habitat, then the species load is certainly acceptable.

One of the possible drawbacks to foxes and cats scavenging around bins at the same times is that one might fall prey to the other. Generally carnivore flesh is not considered tasty to other carnivores. Nonetheless cats have ended up in foxes stomachs. A study in Sweden found cats having been eaten by almost 2 per cent of the foxes examined. Cat remains at a similar frequency have been found in London's foxes. That foxes on occasions will eat a cat seems fairly certain. For instance I know of a convent in South London where the nuns saw from their windows a fox settle down on their lawn to eat a cat, leaving the head and legs as remains.

However, I have yet to come across firm evidence that foxes kill cats. Foxes, like cats, are enterprising scavengers as well as hunters and not averse to carrion. To put this in perspective, foxes also end up in other foxes stomachs, probably as carrion, and removing cubs that do not survive. However, watchmen in country premises will sometimes say they think they have seen foxes wandering off with feral kittens, but these possible sightings are hard to substantiate.

CHAPTER 9
Scent Marking

In an animal like the cat that is so dependent on hearing and sight, it is easy to underestimate the importance of scent and scent marking. Previous undervaluing of the cat's social life has not helped, for scent marking is usually believed to be of particular significance to a social animal. Nonetheless the role and production of scented materials should not be thought of in isolation, for these materials usually have physiological as well as behavioural roles.

The most obvious case of the scent role not being the primary function is in the deposition of droppings or faeces.

Faeces

Observers of the true wild cat (F. silvestris) have differed in their opinions as to whether it buries its droppings or leaves them in a prominent position. One report on the habit of caged wildcats leaving their faeces in their drinking water bowl has even been interpreted as the height of discretion, suggesting the use of mountain streams to wash the material away! However as many species of captive mammals foul water bowls it would seem to be an unreasonable extrapolation of what is apparently hypermarking caused by a restricted space.

Faeces are certainly smelly and in a number of animals, including other cats such as the tiger, they are used in some sort of territorial marking, but do they have that sort of significance in the wildlife of the domestic cat?

Most of the time the domestic cat will bury its droppings under soil. The cat scrapes a hole with its forepaws, and sits on its haunches above the hole. It is under a strong urge to cover for if it is deprived of soil the cat will attempt to pull nearby objects over the hole. Certainly the cat's sense of smell is significant during the covering up, for it gauges the need for further earthing with a few careful close sniffs in pauses between scraping earth back. Nonetheless on certain occasions cats that normally bury the material will make no attempt to do so, leaving it prominently sited.

In a three year study in the late 1970s in Scotland it was found that farm cats usually buried their faeces, as did subordinate feral cats on the uninhabited Monach Islands. However, dominant feral cats on the islands deposited faeces on grass tussocks on trails within their territories.

I find that intact house-cat toms will also leave prominently sited unburied faeces along their peripheral trails, particularly when they are asserting a particular limit that is under dispute.

Such is the volatile nature of smells that fully dry faeces make little smell and the aromatic life of faeces can be extended by burying. The covering may reduce the initial intensity but the longer term smell over many days encompasses an area heightened by bacterial action.

The habitat of some feral factory cats can provide problems to the animals. In foundries cats tend to make use of sand heaps near the hot metal shops, while in clean factories cats may be reduced to defaecating on a small amount of dust at the base of a pillar, or communal latrines may appear behind standing equipment or in cupboards. Although I have not found industrial feral cats invariably establishing communal latrines, nonetheless it is sufficiently common to note. These tend to be, but are not always, in locations where choice of individual latrine sites may be poor.

Domestic cats normally use fresh individual latrine sites, often towards the periphery of their current confident territory – which often results in your neighbour's seed bed being dug up!

A domestic cat does not have an infinite choice of latrine sites as much of its range will be covered by buildings, grass and concrete. The soil of permanent flower beds is often compressed or rain-panned and so hard to dig, so vegetable seed beds are ideal.

Skin scent glands

Man is a nearly nude animal and in the absence of hair is covered with eccrine or the watery sweat glands to control temperature. Cats being furry have no need to resort to sweat glands all over, which would be quite dangerous with that much hair, for the rate of chilling would be considerable. Instead the cat takes great care that it does not overheat by avoiding unnecessary exertion, and carefully controls how much time it sunbathes in direct heat. This does point to those eccrine glands that the cat does have on the pads of its paws having another prime function other than keeping cool feet on a hot tin roof! The role seems to be one of keeping the pads moist so that they do not crack or flake, such that not only can these essential surfaces remain flexible to gain good grip but that their touch discrimination retains its sensitivity. For the night walking cat an instant appreciation of the texture of the surface being walked on is essential. These glands in the cat do produce an aroma detectable even to the human nose on close inspection, but probably less strongly than in other animals. How detectable by another cat such a faint scent footprint would be without a visual guide is doubtful, but cats rake wood and trees (usually semi-peripherally to their territory) and by repeated raking at the same spots for that cat, the wood surface for a foot or so takes on a frayed appearance at which cats will give investigatory sniffs.

The sebum produced by sebaceous glands has a role in protecting the hair and skin from retaining water – something most cats are very fussy about.

In many species of mammal, castration of males with subsequent androgen fall, reduces secretion and smell of skin glands.

Group cats by cross-rubbing and sleep contacts overlay a group scent on their personal scents. The group scent may have an overall identity, but as contact and scent production of individuals within the group differs, the group scent will be a variable composite.

Domestic cats develop scent association with their 'human conspecifics' and much rubbing gains a joint group recognition scent which also varies in content relative to amounts of contact. This association gives reassurance and so jumpers taken off and left on a chair or bed will be sniffed and then 'knitted' by the cat. Cats will sometimes develop a strong attachment and behave towards the apocrine scent from the armpit of their human 'conspecific' in a manner normally reserved for catnip. Ecstatic rubbing and licking will continue for considerable periods if the modesty, alarm or ticklishness of the person involved is not overcome!

One of the fastidious actions of most domestic cats that probably does not endear them to non-cat loving people is the rigorous way in which a cat will methodically wash itself after it has been tickled or stroked. The cat patiently waits or as some put it, tolerates, the attention of the tickler and then washes. This need not to be taken as a slight on the cleanliness of the tickler. As much as washing, the cat is combing the fur from being ruffled back to lie in the correct direction using the papillae on its tongue. It will also be reinforcing its own scent, while at the same time it may be reading by taste the scent of the tickler.

The licking and thus the obtaining of information about scent by direct taste occurs in many animals, and may play a larger role in the cat than is at first apparent. Licking kittens is a fairly clear instance, but others do occur. Sometimes in the 'Flehmen' (see below) response the tongue will flicker onto the nose, and may touch the material. During washing, contact with other cats is perhaps re-investigated by taste.

Sex and scent

Reinforcement by taste or scent may be one of the reasons for the violent washing of the urinogenital region by both sexes directly following mating.

The cat has enlarged sebaceous glands on the chin plus specialised skin glands in other areas. Enlarged skin glands have been found at the root of the tail of the cat but these do not become enlarged until the advent of sexual maturity. These glands are larger in the male than in the female, and become most active at mating time.

Both just inside and outside the anus the skin contains a number of small glands, organised into elaborate structures in some animals. Anal sacs occur in most species of carnivorous mammals and in both the cat and dog family the structure is relatively uncomplicated. In a large cat such as a lion they are about an inch long, but are proportionately smaller in the domestic cat. The sacs are lined by secretory cells and their discharge would seem to be under hormonal control. The sacs open to the skin surface close to and on either side of the cat's anus. Due to this position they may contribute scent to deposited faeces. Certainly I find that female cats' anal scent glands are at their most odourific (a 'cheesy' smell) on a fortnightly cycle corresponding with greater displays of head-rubbing and affection linked with the cat's oestrous pattern.

Scent spreading

The prime sources of smell in the cat are the faeces, urine, anal glands, vaginal area, skin glands, ear canker and saliva. Scents on the animal can be spread over a larger surface of the animal. Much of a cat's washing time is spent sitting with backlegs stretched wide combing in a radial fashion away from the origin. The cat, by repeated washings from the anal-genital region down the first third of the ventral side of the tail and similarly out onto the ventral sides of the thighs of the back legs, produces a greater surface for release of these significant smells. These areas form a containing pocket in a 'normal' walking stance that becomes suddenly exposed in the greeting posture when the tail is whipped erect. When the cat lifts its tail in greeting to another cat (or even when being stroked) a clear message, particularly of the female cat's sexual state, is wafted about.

Whenever a human 'conspecific' enters a house, perhaps with shopping in a box from the supermarket or returning home having stroked another cat from a friend's house, the cat will usually investigate by careful sniffing such boxes, hands and clothes of the homecomers. Scent of outside encounters are noted and signs can occasionally be seen of a jealous cat that needs reassurance.

Scent sticks

Whenever a domestic cat goes out into the garden its first concern is to carefully sniff the nearest protruding scent 'letter-box' or 'scent-stick'. Doorways are recognised by cats including marauding toms for what they are, key centre pathways from which routes branch out. Their significance is recognised even when shut and the paths beyond closed off. As such, within a few yards of a door, a scent-post can be found. The cat will carefully continue smelling a series of such nearby posts for clues of an intruder before it will allow itself to relax on finding all clear. (The term 'scent letter-box' for a scent-stick is probably more indicative of its role than the more usual 'scent sign-post', for the latter implies direction, when its focal nature only reveals who and possibly when, but not in which direction – a number have to be read before the route used becomes clear).

Although the investigation of a scent-stick can on odd occasions develop into a Flehmen response, the more normal event is a careful but brief sniff or two, not often lasting longer than ten seconds or so, before the cat moves onto the next scent-stick, often after re-marking the stick. Cats will do this to discover who has passed by, but as they will also do this when they have followed another cat, only yards behind, it is apparent that further information about the other cat is likely to be obtained by this method. It may be just to reinforce the visual identification or it may be to obtain information on the current state of the other cat, such as anxiety, sexual state, etc. However, for a previously unseen or infrequently seen animal a link must be forged between visual and scent recognition by the observing cat.

The sticks are marked in two distinct ways. The first is by walking lightly against it, often along the side of the body and without any apparent conscious involvement until the route is examined (so much so that a casual human observer is usually oblivious of the event). The second method is more definite where the side of the chin, just below the lower lip, will be rubbed once or twice against the tip of the stick. The upper lip may curl while this happens and the tip of the stick may even rub on the lips.

The scent-stick so marked is usually firmer than the ones just brushed against which will include tips of grass, leaves and shrub leaves. The selection of these marked objects is less casual than might be thought, for the 'reading cat' will make directly for them. They are clear projections into the cat-path.

The marking material arises from the skin glands and is washed over the body surface. Often relatively little direct focal attention is paid by female cats to a specifically sprayed object by a tom (probably as the generally broadcast nature of the aroma makes this unnecessary), but in contrast much attention is given to the scent-sticks. This is particularly so on moving into a new area, or coming into an area that could not be clearly seen from the previous position. Small gaps in fences or cat-flaps and similarly restricted access places gather considerable quantities of grease and hairs from the cat's coat and are often investigated by cats as scent-sticks.

Spraying

Characteristically when marking, a tom will reverse up to a vertical plane and lifting his rump high, with tail erect, will discharge a highly pungent stream of a volume, usually between 0.2.–1.6.ml, against the object to be marked, accompanied by a judder of rump and tail. It is usually thought that a male will spray scent over a rival's mark to show he was last 'in possession' in that area. However, it may be that the tom is merely hypermarking all the key points of an area and that recent aggressive encounters with the adjacent male have stimulated such actions, rather than specifically smelling the rival's mark and overlaying it.

Marking on someone else's scent or remarking your own may also have time function. It is often suggested that an animal detecting the scent marking of an animal previously passing may have some idea of the time lapse involved. These suggestions which imply noting fresh smelling marks from old marks have been questioned recently. It has been suggested that superimposing one's mark over a previous mark tells you whether or not another has passed by. However this presupposes that either the animals 'all play the same game' and scent mark certain recognisable features (which implies that certain features trigger, either genetically or learnt, a response to mark), or that the scent may be the direction drawn to a mark. In either case certain information is likely to be carried such as sex, relative age and sexual status and possibly specific identity.

Although undisputed dominant toms may spray an area to show they were last in possession, uneasiness and conflict certainly accentuate the activity. Some insight may be gained from the cats of Fitzroy Square. On the evening outlined in the Appendix about a year after the colony neutering, an adult male black tom tried to gain admission to the group, particularly at feeding time. His direct approach was not allowed by the group cats, but he hung around in a submissive manner until the group had dispersed for the evening. Then, looking furtively around, he made for the centre of the feeding area which he carefully sprayed and then moved off. It could be that the group scent, as well as acting as membership ticket to an exclusive club, could also be a way that a stray could both put in an application to join and serve a probationary period before acceptance. Being a long established family club, the Fitzroy Square cats do not however admit members merely because they apply, unlike some of the newer clubs where evening dress is not obligatory and all sorts of coat colours are worn, and virtually anyone is allowed in. Whilst this is a reasonable analogy as far as it goes, it should obviously not be pushed too far. A feeling of confidence from ownership and belonging is obviously important to a cat, particularly toms. For example, in the world of cat breeding a male is often unable to mate if he is placed in a new and strange cage and he is not allowed to spray it first to establish a right to be there by precedence.

The modern urban cat's spray telephone system does suffer from unfortunate crossed lines with long distance subscribers! The car figures largely in the urban cat's landscape, convenient for sitting under to watch the world, dry, sheltered, and low enough to offer comparative safety from passing dogs as well as people. The most convenient place to spray a mark is of course the hub-cap, but on later re-investigation due to the mobility of cars, the hub-cap will have been replaced by another with perhaps no scent, or that of an apparent intruder, instantly requiring marking.

Scent marking can be used as an area-assertive weapon as many domestic cat owners are probably only too aware if they have a hyper-aggressive intact male in their neighbourhood. A domestic cat can be afraid to enter part of its normal home range, including its own house if such a dominant male has left his all-invading pungent 'visiting card'.

In such circumstances the anxious cat will only creep in when accompanied by its human 'conspecific' and will then furtively examine the area to verify the tom has left.

It does not seem to be fear of the scent but rather fear by association. Previous to an aggressive encounter the 'victim' cat may have been ambivalent to such marking. However, after an aggressive encounter the tom will seek out and mark 'key paths' and this means doorways and perhaps windows for a house, and if it can find access the feeding area of the 'victim' cat. Such associated conditioning tells the 'victim' cat fresh spraying precedes an attack. This suggestion that the interpretation of scent is not on a one for one basis has its parallel in spoken language where meaning can be changed by intonation, and how and where it is used.

The cat's ability to detect smells is largely governed by the strength of the smell itself and how well it travels. The pervasive smell of a tom's spray can be detected by another cat in still air within the confines of a building at about forty or so feet. Almost needless to say this is subject to a number of variables.

In the male cat castration removes the smell of spraying and indeed is one of the main reasons given by owners of toms for the operation.

The ability to taste – smell (Flehmen)

When sniffing catnip or some cat urine or similarly scented materials the cat can then be seen to enter into a trance-like state while curiously gaping – the 'Flehmen response'.

The German word 'Flehmen' has no true counterpart in English, perhaps the nearest being 'grimacing', but it is hardly a full meaning.

The adoption of the curious gaping pose enables smells to be wafted and retained into such a position that they can be 'taste-smelled'. This response can sometimes be seen when a cat examines a focal scent location on the ground just before a sequence of rolling and in the gaps between rolling, verifying the focal nature of the scent. A full Flehmen response is not usual when a cat is sniffing the scent-posts while walking through its range, but nonetheless it does occur on the odd occasion.

Many animals, including cats, have a curious organ, the vomero-nasal (or Jacobson's) organ over the roof of their mouths. These Jacobson's organs lie on the nasal cavity floor on the sides of the nasal septum and open to the mouth just behind the first incisor teeth. This physical arrangement allows for the taste and smell action when the animal adopts a grimacing posture and sucks air over it. The mouth gapes and sometimes the tongue flickers and breathing is halted for a few seconds while the animal concentrates in a pensive manner. As in most animals one of the scents causing such a response in the domestic cat derives from the urine of conspecifics. (Sex hormones are voided in urine).

Like many primates, we are devoid of such an organ and so can only guess at the sensation induced, but it seems to be strongly linked with sexual behaviour. It is probable that Jacobson's organ is connected to the medial hypothalamus (involved with sexual activity) and to the ventro-medial hypothalamus (concerned with feeding regulation). I have found it frequently used by cats with focal scenting prior to chinning. It is also seen in cats responding to catnip.

Kittens only start to develop the Flehman response from five weeks of age. Before that age, young kittens are upset and distressed by smells they have not yet met, and are calmed by the smell of their mother and the nest area. This change of responsiveness is one of the factors acting as a very advantageous check on the kittens on any tendency to wander until they are ready for the outside world.

Catnip and catpeople

Most cat owners are aware that catnip exerts a powerful attraction over some cats. They are not alone, for any gardeners trying to establish catnip in their herb garden can be recognised by their sorrowful mutterings over the residual leaf stumps following interest by the local cat. Certain scents bring on particular interest in cats, such as catnip where effusive rolling, sniffing and biting of the plant occur. Nonetheless most people would put this effect under such a heading as 'interesting phenomenon' though not of profound importance, but there is more to it than that.

Catnip (Nepeta cataria) belongs to the mint family and is found widely distributed in the wild state in America and Europe. It is not just the domestic cat that goes crazy over catnip, for even fully grown African lions behave in the same way. This behaviour is very similar to female oestrous behaviour. Consequently, the catnip smell which causes the effect (an unsaturated lactone, trans-cis hepetalactone) might resemble a sexual odour.

It may be that the smell catnip gives off is one involved in the sexual cycle of the female cat, due to the intensity of the identical response by the cats. Another plant (Actinidia polygama) used by Japanese researchers which also includes a lactone in the active components, produced a hyper-response compared to the catnip extract, giving the impression that if the components are mimicking the normal cat oestrous smells, then these may be a super-stimulus (i.e. more active than the cat's own sexual odours).

The effect of catnip is not seen in all cats, but in those where it does occur it is inherited. An extended chin rubbing is seen, with shoulder-rolls, rather than head-over rolls, and body-rubbing. A single-minded intensity overtakes the cat during this activity.

Whatever the sex and whether a cat is sexually intact or not has no effect, as the reaction is found to be the same in neutered and intact cats, both male and female. Both kittens and adults alike demonstrate the behaviour.

The normal parallel occurrence of single-minded rolling happens in oestrus and following copulation. It has been suggested that under oestrogen the skin sensitivity of the chin and head changes and that catnip causes a similar sensation. It is possible that the hormone could affect the facial glandular secretions giving rise to an increase in rubbing and thus to a deposition of scent material. It may be that such 'working through' of part of a behaviour pattern elicits the single-minded intensity in the rubbing.

I have seen the feral cats in the grassed Fitzroy Square study react to focal positions where another cat had sat. Between chinning sequences the focal nature of the spot was checked by the cat. Toms *tend* not to spray on flat surfaces, so it may be that such focal attention to where a cat had previously sat might be in response to the smell of the anal scent glands.

Response to scent

Two distinct types of response seem to occur to a key smell – the Flehmen response, and head rubbing – rolling and chinning on horizontal surfaces. Flehmen seems normally to occur in response to smell from urine or faeces of a conspecific. I have found on occasions that the Flehmen response can accompany the other response.

The rubbing can be carried out in a matter of fact manner, with pauses while the animal looks around. In other circumstances the rubbing will be more determined, more focal, and the animal will be obsessively determined, with body flat to the ground. In the final case the obsessive sniffing and rubbing becomes so intense that the typical Flehmen response becomes an integral part of the sniffing.

The initiating smell that stimulates the rubbing as Paul Leyhausen noted seems to be less specific and of a wider range. Leyhausen found that the surface so rubbed during chinning often becomes quite wet from saliva and in some part from the secretion of the chin and cheek glands. A number of individual hairs from the cat's head also become visibly stuck onto the rubbing site.

With this sort of a response the initial scent may be of less import than the increasingly overlaid scent from the combined saliva and scent of the animal involved. The build-up of response to an ecstatic state seems to depend on the mental and hormonal state of the cat. The same rubbing and rolling is typically seen in oestrous and in pre-copulatory behaviour by female cats and when really intense it triggers greater interest in the males. The role it may be playing is a 'masturbatory' one, preparing the female cat for mounting. In oestrous the cat, being under hormonal control, finds that this self-stimulation is obligatory as her mood is changed. In this instance her chinning has no need to be initiated from without by a low key stimulus, and the rubbing can start without a focus. At other times a low key scent may trigger the cat's memory and once the rubbing starts the intensity, if it increases, will probably be reinforced by the cat's own rubbed scent and by the cat's own increased involvement in a 'masturbatory' way. The initial scent possibly initiates the self-stimulating act during which the action becomes the true focus.

The essential part of the mood of the cat can be seen when following such a non-oestrous chinning episode the cat is re-offered an object that initiated the sequence and totally ignores it.

CHAPTER 10
How Big is a Cat?

There are a series of widely differing stories over what happens to the domestic cat when it pursues a feral lifestyle. Some suggest a reversion towards the believed wildtype similar to a Scottish wild cat. Some say that the animals become larger and more fierce as an adaptation to a more self dependent life. In this vein a number of news reports of feral cats have emerged from south-east Australia where 'Giant wild cats weighing up to 25lbs are roaming central Australia and wiping out thousands of native animals and birds. Most weigh about 15lbs – with some up to 25lbs', so reported the *Sydney Morning Herald* in 1971. Following on these the mythology of the area was increased by the publication of a gripping adventure story built around the presence of the enlarged feral cats.

Fears have been expressed among animal welfare organisation members in Britain that the 'poor little pussies' must be starving to death and very thin. This seems to be based on the premise that because they were domestic animals they cannot possibly look after themselves. The large number of feral cats existing would seem to suggest they can look after themselves very successfully.

Two clear parameters of size exist to enable comparisons – weight and size.

Weight

How heavy first of all are our native Scottish wild cats?

	Sex	*Mean Wt.*	*Range*
Up to 1940	♂	11.3lb	6½–15½lb
After 1940	♂	10.3lb	
	♀	8.5lb	5–12½lb

The slightly lighter mean weights of the more recent figures may reflect hybridisation of the wild cat population with feral cats.

Something of an unfair comparison can be made by looking at the highest recorded weight obtained by a domestic cat which was by a nine year old tom in Connecticut, USA who reached 43lb. The largest domestic cat recorded in Britain reached 42lb. Domestic cats that become overtly obese normally do so due to a hormone imbalance such as hypothyroidism. One cat I knew in such a distressing condition was so round as to be almost spherical, the four paws seemingly tacked on to give corners to the football shape were unable to reach the ground all at the same time.

To find a normal range as encountered in the neutered domestic population I weighed a group of adult cats living in comparable conditions and given free access to food.

Sex	*Adult Mean Wt.*	*Range*
♂	7.7lb	4½–10lb
♀	6.1lb	4–8lb

The neutered domestic house cat is normally a smaller animal than the Scottish wild cat.

The neutered cat can on occasions put on a little weight in the form of fat. The intact tom may also put on weight, but this is more likely to be in the form of solid muscle. I have weighed large domestic intact toms in the East London study area (Chapter 5) and found them to be around 14lb. Similarly a semi-farm cat I knew well in Essex, who was largely self-supporting, also maintained the same weight of 14lb for much of his life.

I have weighed the intact, totally free-living feral cats from seven colonies in England, with the following results:

Sex	*Adult Mean Wt.*	*Range*
♂	9.1lb	6–11lb
♀	5.7lb	3½–7½lb

When these are all put together as a guide for comparison of their mean values it seems that the feral cat is certainly not underweight when compared to domestic cats. As in my north Romford study area intact toms were 18 per cent of the population, the neutered figure would need amending to become a normal domestic figure, but even allowing for this the male feral cat mean remains a higher value.

The females give a relatively constant baseline for their type. Both neutered domestic and feral females are about the same weight, and both before and after 1940 wild cat females stayed constant. The mean weight shifts occur in the male, with the feral and wild cat approaching each other. The feral life may toughen up toms, but hybridising with feral cats seems to be reducing the wild cat toms.

Length

A glance at the mean size of any of the parameters that I measured on the domestic cats shows that the average male is longer in each of these respects than the average female.

When the same parameters were checked for the feral population, the same difference also held for the feral animals.

However, the average feral female cat was smaller in all measured parameters than the domestic female cat except in the diameter of the abdomen and length of the head.

Similarly, the average feral male cat was smaller in all measurements than the male domestic cat except again in the abdominal diameter and the head length with the telling additional exception of the back leg. However, for the male the differences are so small as to be insignificant, the larger mean differences being seen for the females, which is the reverse condition to that found for body weights.

The larger feral abdomen may reflect poorer diet but generally the lengths similar to the weights, suggest that the average English feral cat is neither a huge monster, not yet a particularly undersized, underfed individual in present circumstances, and hardly different from the average domestic cat. Occasionally huge muscled 'monster' feral cats do turn up and may be the origins of media-mouthed puma sightings, but similarly huge domestic cats can be found, although these may be obese rather than muscled! Despite these, the normally encountered feral cat appears so similar to the domestic cat that differences must be sought in terms of behaviour, home ranges and factors such as diet.

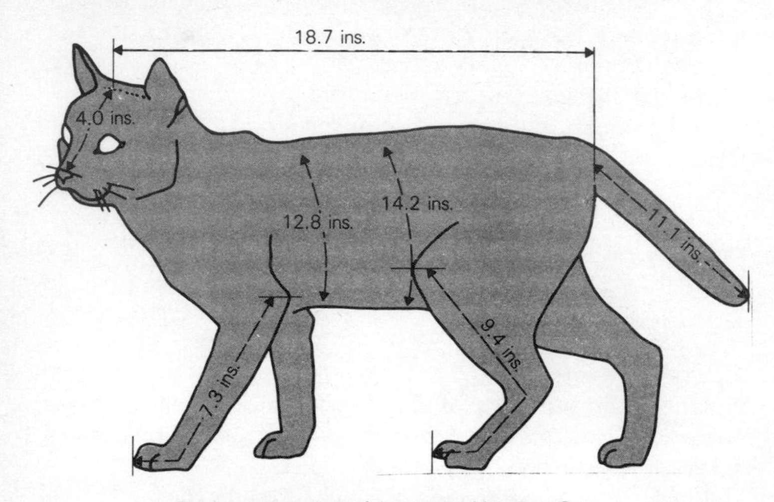

The linear dimensions of the average Male Feral Cat

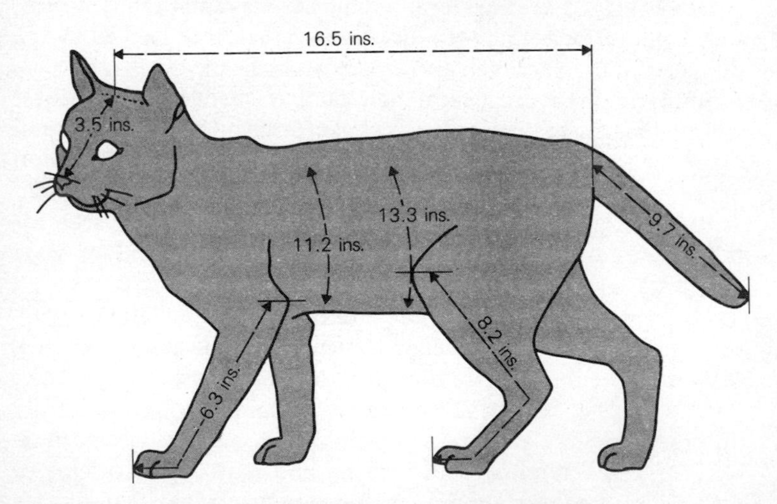

The linear dimensions of the average Female Feral Cat

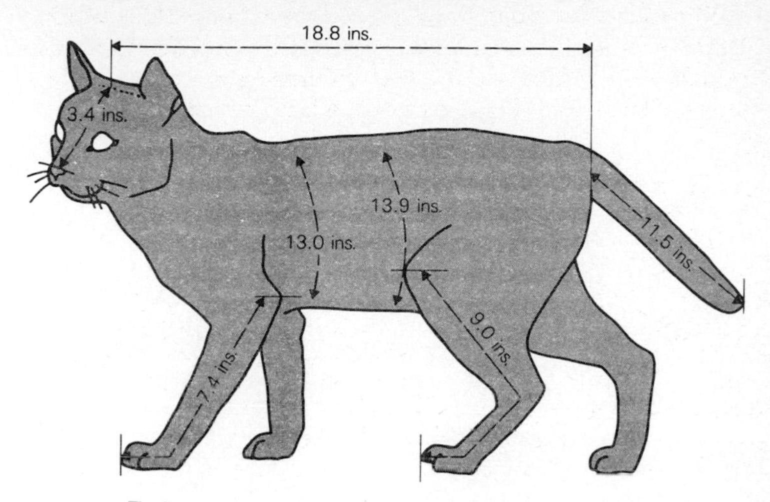

The linear dimensions of the average Male Neutered House Cat

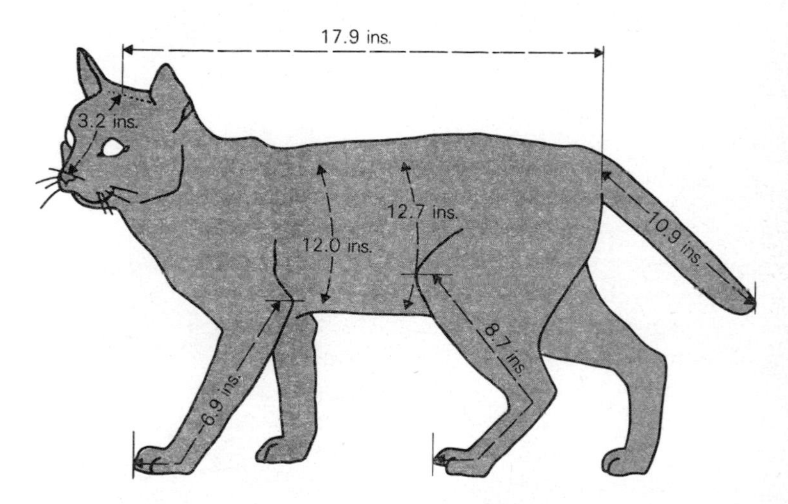

The linear dimensions of the average Female Neutered House Cat

Comparison of sizes between feral and neutered house cats in England.

CHAPTER 11

Feral Cats and Islands

Feral cats are found all over the world in many habitats. In most countries they can be found in both rural and urban conditions. Frequently they have become among the dominant predators on small islands dotted well across the globe from the equator to near the antarctic. On uninhabited islands the cats cannot scavenge from man and although they will eat carrion, must hunt to survive. A notable feature of small islands is the paucity of species in the fauna. The fauna that is long established by being isolated can evolve on separate routes from the mainland stock.

The classic example of this, on the equator in the Pacific off South America, is the group of Galapagos Islands. On Darwin's brief visit to the islands in 1856 he had drawn to his attention the slight variation between the islands' finches and giant tortoises that set his mind musing on a mechanism to produce this. The musings led to the crystallisation of his theory to rationalise these differences, by natural selection making evolution work.

The indigenous or native evolved wildlife face many problems since man discovered the islands and introduced both accidentally and intentionally more competitive animals. The slower established unique reptiles have been fighting a losing battle against the invaders. One of the introduced mammals is the cat, many surviving in the feral state.

Feral cats, along with feral pigs and dogs, became a major direct threat to the giant tortoises on the islands. Fortunately, a breeding programme of each island's unique tortoises at the Charles Darwin Research Station on the Galapagos has turned the tide. The islands are a national park of Equador, and the warden's report that on San Cristobal island where a watch has been kept, the cats are found even in the most arid and inhospitable parts, using crevices in the lava formed rocks for shelter. They are active day and night and are the principal cause of the decline of frigate birds on the island. There has been 60 per cent chick mortality of not only the two species of frigate bird, but also of the three species of boobies (southern gannets) and of the unique swallow-tailed gull on the island, and the cats are high on the list of suspects.

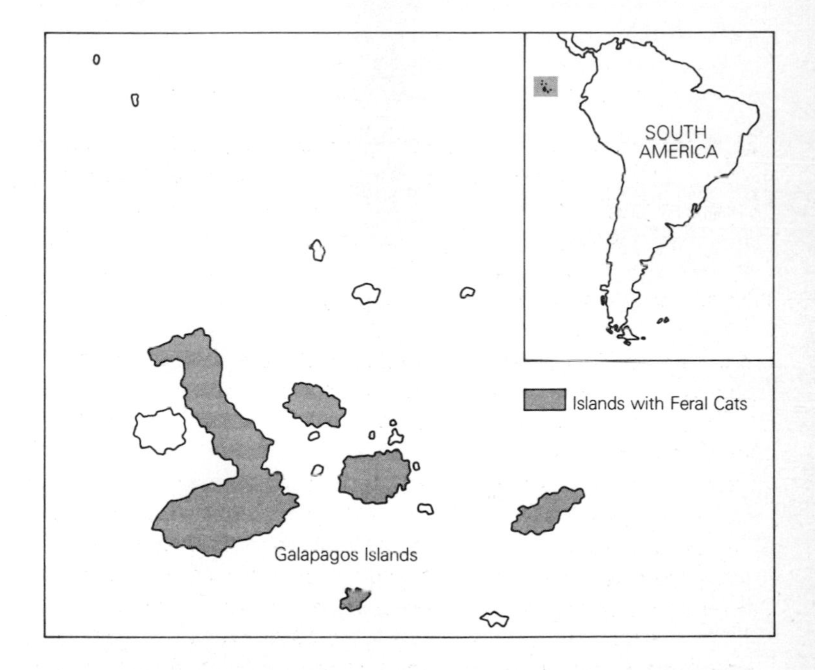

The Galapagos Islands lie on the Equator some 600 miles off the coast of Ecuador, South America. Feral cats arrived with man, and so far are only on some of the islands.

The cats may have a density on another Galapagos Island, Santa Cruz of up to 60 per sq. mile although as the area investigated contained rare fresh water, this is probably higher than the islands can support overall. Nonetheless they are at quite high densities all over the island. The cats have been found to be extremely bold, entering the large sea-birds' nests with the adults present. They have a particular liking for young marine iguanas and their relatives the small lava lizards, but have grasshoppers as their staple diet.

The Galapagos are particularly noted for Darwin's finches and like cats anywhere the cats on the Galapagos like small birds and are particularly efficient ambush predators of these ground dwelling finches.

Kittens on the islands when seeing a stranger become very quiet and melt away into the brush. The Galapagos Island's unique wildlife takes some revenge however in that the Galapagos Hawk successfully preys on the feral cats' kittens.

The cat has gained near world-wide distribution by association with man, including to the inhospitable sub-antarctic islands, where the remarkable cat survives the extreme conditions in a feral state.

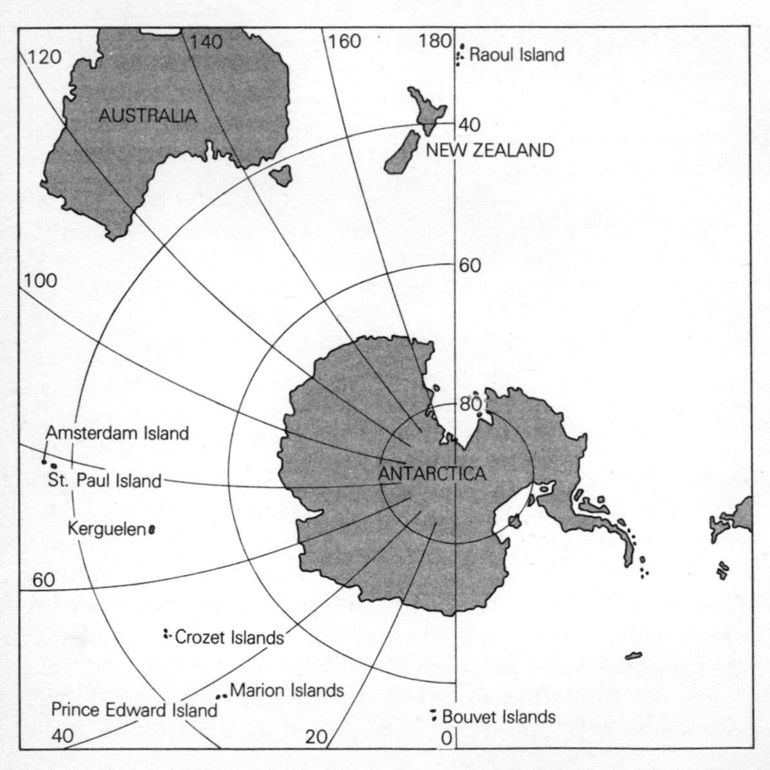

Another group of volcanically formed islands at the other climatic extreme are the Marion and Prince Edward Islands, which have a mean temperature of 5°C and are lashed by gale force winds, bringing an annual 260cm. water as rain, snow and hail. In consequence the flora is typically Sub-Antarctic, being low in height. The situation of introduced mammals is more simple on Marion Island than on the Galapagos. On Marion Island there are only two introduced mammals – the cat and the mouse. A simple attempt at biological control was carried out in 1949 by the introduction of five cats at the Meteorological Station to control the mice that are presumed to have landed from occasional visiting sealing ships. Estimates are that there had been an increase to somewhere between 500 and 2000 cats on Marion Island by 1974. Other Sub-Antarctic islands, particularly the Kerguelen and Crozet groups and New Amsterdam Island, have also suffered invasion by this hardy duo, the cat and mouse. On the Kerguelen Islands the density has been estimated to be five adult cats per hectare.

However, the cats are not host specific predators and despite the intentions of the introducers, heavily supplemented their diet with burrowing petrels. One of the Marion Island cats has even been seen to kill an adult Macaroni penguin. However, more usually, eating of both penguin and seal carcases is as carrion. In the desolate landscape of these islands humps and fissures of the lava provide some protection and the cats make lairs among these.

Ascension Island

On Ascension Island the Wideawake or Sooty Terns undergo a long period before being ready for egg laying or spending the night on the breeding grounds, and the day back at the colony. Until the introduction of the cat in about 1820 there were no night predators and the cat dived straight into exploiting this advantageous niche, and continues to do so among the present three-quarters of a million Sooty Terns.

In 1979 came news that cats were perhaps destroying, aided by rats, the Sooty Tern colony at Denham Bat, on Raoul Island, of the Kermadec Group, to the north of New Zealand.

On the dry and fairly barren Ascension Island there are few rats. Despite cats previously eradicating the once large breeding populations of frigate birds, boobies, noddy terns, tropic birds and petrels from any accessible part of Ascension Island, the population of Sooty Terns continued to thrive. Their large numbers and breeding together may be the secret of their success and the virtual complete absence of food for three months rigorously checking the cat population while they are away.

On Raoul Island the other seabird colonies have similarly been destroyed, but the Sooty Tern population is also falling. There the rats are more abundant and as well as directly attacking the chicks, they also provide an alternative food source for the cats while the terns are away.

The wildlife of New Zealand's groups of islands has suffered many predations from cats. The Bush and Rock Wrens (New Zealand Group Wrens) have been decimated by introduced cats – with help from introduced rats and stoats. More particular was the case of the Stephens Island Rock Wren where the lighthouse keeper's cat exterminated the entire population. However, the cat took its catches home, otherwise we would have been unaware that this particular species of New Zealand Wren had ever existed.

CHAPTER 12
Health and Safety Cat

Cats and their passengers

Life for a parasite within a host may be a leisurely one, bathing in food with relative security from predators, but once outside the host the world could come as a rude shock. To minimise this the stages of parasite development can occur in specific hosts. This means that some insight can be obtained into what the cat has been living on from the parasites in its gut or digestive system. For example a tapeworm (or cestode) called Taenia taeniaformis uses small rodents such as mice and rats as its intermediate host. Cats with this tapeworm have therefore been eating rodents and a knowledge of distribution of Taenia taeniaformis in cats will help to establish how efficient they are in controlling the rodent population.

Parasitoligost colleagues are examining the feral cat intestinal parasite load for London and the surrounding thirty-five mile radius. Needless to say this is a long term project. The results, however, are helping to evaluate properly the feral cat's role as urban predator. The findings couple with the stomach contents analysis to shed light on the survival diet of the urban feral cat and therefore upon the cats behaviour. Of course, caution must be used in interpretation, for although a high value for Taenia taeniaformis means a high intake of rodents, a low incidence does not necessarily mean the reverse, for there could also be a low level of parasite in the intermediate host.

In Britain there are two tapeworms that commonly live in the cat's intestine and they are the Taenia taeniaformis and Dipylidium canium. According to Sloan these are detectable in cat faeces examined in veterinary practice at 3 per cent and 2 per cent respectively. Similarly two roundworms (nematodes) commonly occur in the cat intestine, these being Toxocara cati and Toxiascaris leonina, which according to Sloan occur at 20 per cent and under 1 per cent respectively.

In Britain these intestinal parasites give little clinical concern to the vet for the health of the animal. Indeed, I have found that in the rural feral cat, tapeworms are often found in the fit, magnificently large toms that are supporting themselves by catching rodents. The rodent provides the best balanced diet to the cat and it would seem that the rugged fitness derived from the best diet, plenty of exercise and interest in life more than redresses the loss of nutrient to the food poacher in the gut. (The tapeworm in the photograph section came from a feral tabby tom who weighed 9½lb and was 18½ inches long with a chest of 12½ inches. He was a very successful rodent hunter.)

While Taenia taeniaformis results from catching infected rodents, Dipylidium canium does not result from hunting activities, but derives from the vectors or intermediate hosts, the cat flea and louse. The Dipylidium canium causes the most concern to cat owners, although its presence seems to have little effect on the cat. The proglottids or mature segments break off the end of the worm to restart the life cycle of the parasite. However they are mobile, and the sight of these off-white 'things' that have all the appearance of flatworms crawling out of the cat's anus and into the fur is disturbing to most people. Equally disturbing to the cat owner is that the proglottid then dries to an appearance identical to a grain of rice, which then falls off or is washed off by the cat, at best onto your favourite velvet-covered armchair, and at worst onto the lap of a visitor wearing a dark suit. One grain of 'rice' would not be so bad, but as luck would have it, on occasions like these the cat settles down well and has a good wash, knocking off some dozen or so, along with the black specks of flea droppings. For with the linked life cycle these two go together. The pale proglottids and the dark flea droppings make sure that whatever the colour of the clothes your visitor notices them. Politeness probably restrains your guest, who perhaps does not like cats anyway (for cats often sit on visitors who don't like them – for such people are less likely to fuss them about and will let them settle comfortably). However restraint disappears when a huge flea visibly runs across the cheek or lower jaw of your cat!

A few experiences like this and the most nonchalant owner is knocking on the door of the vet. While the vet can remove the tapeworm with arecoline acetarsol, or dichlorophen, the life cycle has to be stopped by removal of the fleas or it will recur – and that is less easy. Dusting the cat with a proprietary powder containing an insecticide helps. Some cats find being 'puffed' at – especially with a powder that irritates the nose – unnerving, and will bolt. In such cases the secret is not to 'puff'. Tip or 'puff' the powder to load up a brush and then, holding the cat firmly, gently work it into the coat. However, this will be the least of your worries if the fleas have become established in your deep-pile carpeted house. At one time, with the spread of the vacuum cleaner it seemed that the days of the cat flea, like the UK human flea, were numbered, but with the tide of deep-pile carpets awash from wall to wall, nurtured by central heating, the cat flea revival seems beyond revivalist meetings – and closer to epidemic proportions. Then you have to treat your whole house like your cat's coat.

With high numbers of cat parasites, cats *can* suffer a loss of condition and become listless. Dipylidium canium especially highlights the problems of extrapolating our approach from the domestic cat to the feral cat without first checking our motives. It is easy to think from – 'our fireside cat gets treated if it picks up tapeworms' – to – 'poor gone-wild cats, if they have tapeworms they must be ill (and/or suffering) and something must be done because of this'. This overlooks the largely cosmetic need of humans in controlling cat cestodes, which is not a basic survival priority of the feral cat.

There is an improbably slight risk of young children picking up Dipylidium canium by catching infected fleas and swallowing them. However the medical profession so far has not reeled under a flood or even a trickle of cases. The nematode gut parasites, on the other hand, do present a greater health risk to humans than to cats. This is basically due to parasitic nematodes being very host specific, and if the eggs of Toxocara cati from cats, or Toxocara canis from dogs, are eaten by people the larvae, when they form, cannot follow their normal pattern and wander about the body. The condition is referred to as 'visceral larva migrans'. As the larvae work their way through the organs they can crawl through the eye and brain, and have on occasions caused blindness and even death. As children are not particularly hygienic they could be prone to this hazard. Fortunately the habits of cats are unlike those of children, and it has been thought that the fastidious behaviour in burying faeces greatly reduces the risk of communicating the parasite (although recently doubts have been raised over this). As with the tapeworm, and as feral cats are unlikely to come close to people, they are unlikely to pass on T. cati. Similarly, even domestic cats, far more so than dogs, are likely to remain aloof from children. However, even the dog need not be avoided like the plague, for the main danger time lies with puppies and nursing bitches and as the dogs age their nematode burden reduces.

If there are only one or two nematodes in the cat stomach they will usually lie against the mucosal wall of the body of the stomach. However, if there are half a dozen or more they tend to congregate in convoluted clusters forming a loose lump into the lumen at the transition of body to pylorus, usually up against a hairball (if present) that invariably is jammed into the pyloric region, wedged against the pyloric sphincter (the emptying muscle of the stomach). It is probably the obstructive nature of the accumulated matted fur ball that causes the cat to reach and vomit it free. But as the nematodes pack against the hairball obstruction the cat will often vomit these free as well. So it is as well to glance occasionally, however, quickly, at the vomited hairball of the household cat as you scrape it up to dispose of it from your favourite carpet! Then run for the piperazine. Don't forget that the dosage rate for an adult cat may well kill a kitten. When in doubt consult your vet.

A group of studies were carried out in south-east Australia on the diet and intestinal parasite load of feral cats. The feral cat has become in that part of the world one of the main predatory carnivores, alongside foxes and wild dogs.

The gone-wild cats roaming the outback of Australia, as their stomach content analysis shows, are supporting themselves on a higher proportion of free-living prey than are most of the feral cats in England. This is perhaps not unexpected with the greater distances between human habitations from which to scavenge in

Australia. An Adelaide hunter, Bill Hambly-Clark, was reported by the *Sydney Morning Herald* in 1971 as saying that the feral cats were wiping out thousands of native animals and birds. He believed that the disappearance of small ground birds from areas was due to the cats and that the cats were also preying heavily on lizards and frogs. That they do eat the reptiles and amphibians can be seen from their internal parasites.

The story of Australian wildlife is in many ways the story of the Galapagos Islands, only on a larger scale. In Australia the isolated group of animals, the Marsupials (non-placental mammals) have evolved to fit ecological niches held by placental mammals on other large continents. This occurred to the extent that a few species of marsupials so resembled cats in their ferocious hunting of small mammals and the ease with which they can run up trees that they were called 'native cats' by settlers. As with the Galapagos, with the coming of western man in ships, in went dogs, rabbits, rats, bats, mice – and cats. Fortunately for the marsupials the size of Australia gave some measure of protection and life is settling down to an uneasy balance.

South-east Australia is the most temperate part of the continent and is the area most populated by man. Tasmania, south of the Bass Strait, is even cooler and wetter.

The incidence of the rodent-mediated tapeworm Taenia taeniaformis in Sydney's stray cats (that ended up in 'cat-pounds' and were put down after a period in which they were not claimed) was lower than for New South Wales. Perhaps this is to be anticipated from our knowledge from London that urban feral and stray cats have a higher proportion of ancillary feeding and scavenged food in their diet than do rural feral cats that generally catch a greater percentage of living prey.

The midland Tasmania level was much lower than King Island. However when the stomach contents of the cats were examined, rodents were found to occur in only 6 per cent of the Tasmanian feral cats, but in 33 per cent of the King Island cats (which seems to offer the explanation for the difference). However Victoria had a huge incidence of T. taeniaformis in its feral cats and a mean number of ten per cat, with the highest having an incredible 102 in its intestines. (This compares with the mean number per cat for New South Wales being two and the highest number per cat being

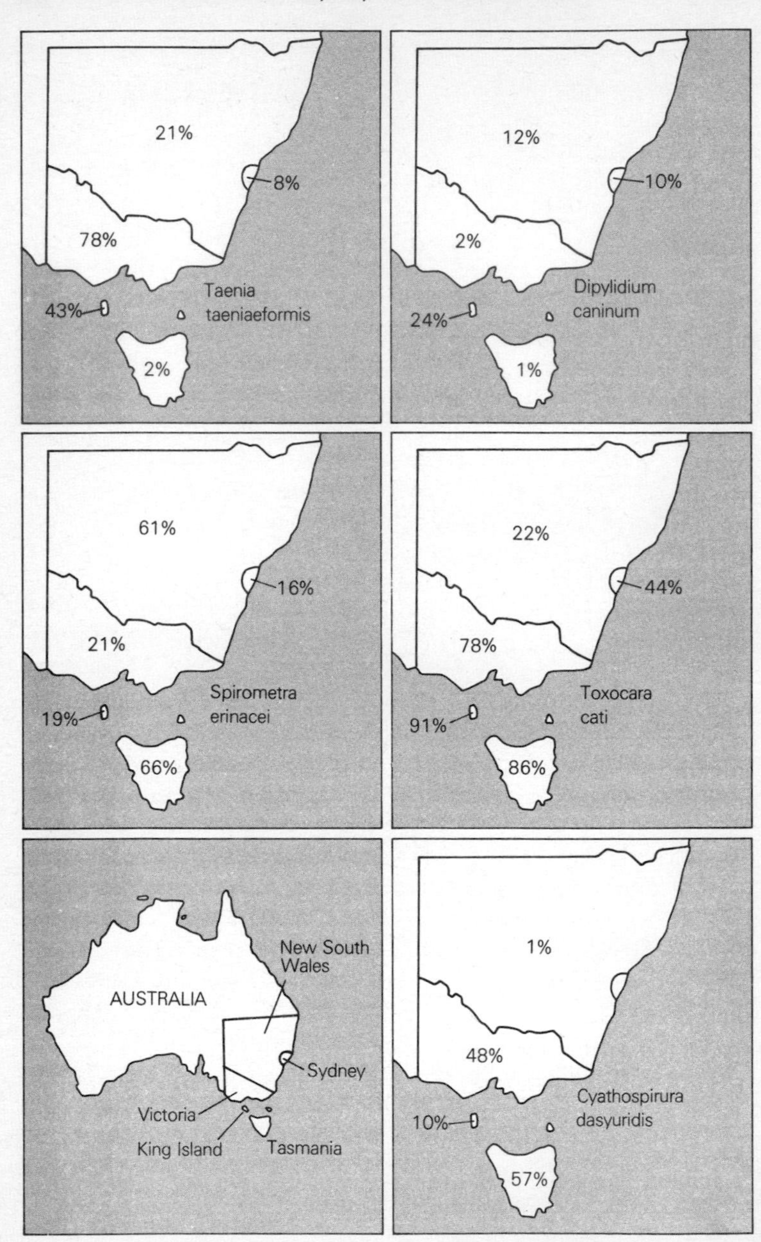
21%
8%
78%
Taenia
taeniaeformis
43%
2%
12%
10%
2%
Dipylidium
caninum
24%
1%
61%
16%
21%
Spirometra
erinacei
19%
66%
22%
44%
78%
Toxocara
cati
91%
86%
New South
Wales
AUSTRALIA
Sydney
Victoria
King Island
Tasmania
1%
48%
Cyathospirura
dasyuridis
10%
57%

ten.) Why should Victoria's cats have had such a high level of T. taeniaformis? In the years 1960–70 of the study there was a mouse plague and from these figures it appears that the feral cat population took considerable advantage of this increased food source. Even so, it kept up its high intake of rabbit as well.

The distribution of the nematode Toxocara cati had a similarly interesting pattern in that the number of cats infected increased as conditions became cooler and wetter. Sydney's cats on the West Coast had more than those in New South Wales. Victoria had more, while King Island and Tasmania were both still higher. This seemed also true for the burden each cat carried, for the mean number per cat in New South Wales was six, while in Victoria it was twelve, and the highest number per cat in New South Wales was twenty-five in Victoria it was 155.

Successive litters of cats were often reared in the same general location, increasing the risk to the kittens of infection with T. cati. Nonetheless cats make very hygienic mothers and do not allow kittens to soil the nest. However there does seem to be a link with T. cati at the kitten stage in feral cats. While I have recorded low infection with T. cati in some colonies, in some colonies both in London and in rural cats as far north as Blackpool about 60 per cent of the cats are infected. That near-adjacent London colonies can have noticeably different T. cati loads reinforces the discrete nature of group movements. The kittens, while probably picking up infection directly from the infected mother, are probably also exposed as Coman suggests to some infection from the re-used immediate area.

Cats' internal parasites can give evidence as to what sort of animals the cats have been eating. Cats with the tapeworm Taenia taenieaformis have been successfully catching and eating rodents, while those with tapeworm Spirometra erinacei have certainly been living off some of Australia's reptiles or amphibians. Other parasites, like the tapeworm Dipylidium caninum and the roundworm Toxocara cati do not show prey, but reflect the cats' way of life. The roundworm Cyathospirura dasyridis is a rare example of a parasite living in the introduced feral cats just as if they were the marsupial 'native cats'. (From the data of Bearup, Coman, Coman and Brunner, Gregory and Munday, and Ryan.)

Australia's reptiles and amphibians carry the cestode Spriro-metra erinacaei and the frequency of its occurrence in feral cats strongly suggests that these form common prey to the cats. Victoria and King Island cats have a lower incidence of the parasite than Tasmania, but then they also have a lower incidence in reptiles and amphibians in stomach contents than does Tasma-nia.

Australia's own 'native cats' Dasyurops and Dasyurus have a nematode of their own, Cyathospiruta dasyuridis. While as yet it has not been found ending up instead in domestic or farm cats, it was found in these studies of feral cats, understandably most in Tasmania. It seems to be one of the limited examples of a parasite of the native marsupials infecting the introduced placental mam-mals. This could be seen as the feral cat really becoming part of south-east Australia's wildlife – or the revenge of the marsupials – depending on one's outlook.

Rabies

At present island Britain is anxiously watching the march of rabies across Europe, fearful that the protection of our water skirt will be breached. It seems only a matter of time before rabies is intro-duced and for all our vigilance most people expect it to be smuggled in by a selfish pet owner on a private yacht, landing in one of England's many small ports and inlets. In all probability it will be a smuggled cat. Cases occur from time to time and are rounded upon.

In the event the authorities should act considerably faster than in the case of Dutch Elm Disease in containment. The immediate fear of some pet owners will be that their animal will be slaught-ered needlessly because it is in an area near an outbreak.

However, if rabies were to become established our carefree approach to pets and wildlife (and farm animals) would change overnight. However, even if it does come across should we become agoraphobic? Rabies in humans has been invariably fatal, but nonetheless they are not dropping down like flies in Europe or anywhere else where rabies has become established.

It is as well to gain some perspective from somewhere that has developed rabies in recent times. Rabies was first diagnosed in wildlife in Canada in 1947, but the first case in Ontario, Canada, was in a fox in 1954. The incidence of rabies in Ontario reached a massive peak in the winter of 1958, and then fell equally massively over the next two summers, after which it maintained stable limits (about 100 cases per month). The seasonal fluctuation continued, up in the winter, down in the summer. It was noted that this general pattern in most animals was preceded by the peak for the fox implying the fox was the main propagator. This fluctuation has also been noted in Czechoslovakia, Denmark and Germany. The fox certainly had more cases than any other animal and apart from the skunk, there were few other wildlife species cases. A high percentage of the foxes unusually attacked porcupines, but no porcupine developed rabies. Perhaps we could anticipate the same attack on hedgehogs due to night meetings. We can certainly anticipate night meetings with cats around dustbins, and such meetings may be higher due to our peculiar situation of suburban foxes. Similarly it might be supposed human contact would also be greater for the same reason.

Nonetheless historically when man has been infected it has been most commonly by the dog, and epidemics in urban dogs were of great concern in Europe and in Virginia and North Carolina in America in the eighteenth century. In the Ontario study it can be seen that the dog just leads the cat in the number of cases with rabies (yet these are astonishingly led by cattle and the thought of a rabid cow is daunting!)

It would seem that rabies follows the same natural growth curve as any other organism, and settles down after an initial high level to a relatively stable and much lower number of cases per year. If rabies arrives here it is to be expected that it will follow a similar pattern. Once established, it would not necessarily ravage through cities and its nearness need not lead to panic. Historically humans have received it from dog bites, but against this the number and size of dogs make the problem of elimination of stray dogs in times of crisis a relatively simple matter. Such controls of stray dogs eradicated rabies from Britain by 1903. In terms of control the number of stray and feral cats is considerably greater than dogs and the cats smaller size and greater agility make them a far greater problem. Similarly if you ever do tangle directly with a rabid cat, the chances of your escaping without your skin being broken are remote. London's present apparently unique large number of urban foxes present a chillingly high probability of contact with urban cats. It may be in this that if rabies comes we shall present a slightly different pattern to the rest of the world. Let's hope the defences hold.

When all confirmed cases of rabies throughout Europe from 1977–9 were looked at, 4–5 per cent were in cats.

The cat usually shows the 'furious form' of rabies, with a prolonged excitement stage, but short paralytic stage. With rabies in a cat all that may be noticed before the excitement state, is a change of temperament or mood. During the excitement stage any sudden noise may trigger bouts of scratching and biting, with obvious salivation. Onset of progressive paralysis and collapse is fairly quickly followed by death. This entire sequence may last anything from two to ten days.

Life expectancy of the cat

Given both health and safety a cat can expect to reach a good age. However, what is the normal life expectancy of a cat, and what is likely to affect it? Cats seem to have the capability for longer lives than dogs. Some cats reach a great age, but this is far from saying that this is normal expectancy.

Documentary evidence exists for a cat called 'Ma' living in Devon to have reached thirty-four years, but the '*Guinness Book of Animal Facts and Feats*' also lists less well accredited claims of up to forty-three years old. More frequently, with firm evidence, cats reach the grand old age of twenty years. Sixteen cats over the age of twenty were tracked down in 1940 in the USA, while in 1956 Alex Comfort found in Britain ten such cats that he believed had definitely reached over twenty years.

One of the major factors affecting the longevity of cats is their sexual status. Female cats, like women, have a head start in the longevity stakes. Consequently spaying female cats does not greatly increase their lives. However, those who do not like the thought of having their tom castrated can take comfort from the great increase in life expectancy he gains. Many intact males suffer demise through fighting, but the longer lives of castrated males were not due just to a less violent life. Castrated males for example were found to be less likely to suffer death as a result of infection.

Neutered cats survive longer than intact cats. An American study in Pennsylvania carefully followed the lives of 669 'moggies' (outbred cats) and found that neutering had extended the mean age of death; although only very slightly for queens, it did make a considerable difference for toms. (The mean neutering age was six months, and most were neutered under one year.)

(A parallel study with breed cats found not only did neutering similarly extend their lives, but that breed cats lacking the hybrid vigor of 'moggies' died noticeably younger (both intact and as neuters).

It is sometimes suggested that intact toms will batter neutered toms just for the sake of so doing. On investigation I find such 'neuter-battering' usually turns out to be from the hyper-aggressive intact males that are occasionally encountered attacking viciously anything feline on sight – intact, neutered, male or female – and even very occasionally people.

Cats castrated before they reached sexual maturity, when compared to those operated on later in life have been generally found to die at an older age. Female cats seem to avoid dying of pyometra or breast cancers with spaying, the latter more so if spayed when young.

One decisive drawback encountered through research has been a higher incidence of cancer in the castrated cat than the intact cat.

In the course of daily life the cat receives acute injuries. The head and extremities have been found to be consistently vulnerable. The cat's jaw is particularly prone to damage by falling or from motor vehicle impact. The cat, on falling from a great height may land 'safely' on its paws, but its jaw may bounce on the ground. The cat may be able to land on its feet from a fall and survive proportionally greater falls than man, but as can be seen that does not mean it does not receive bad injuries, especially from high falls. James Alcock has estimated that he sees, as a working vet, fifty broken cat jaws for every broken dog jaw, largely due to traffic accidents.

The higher damage to the abdomen of cats than the thorax or neck of cats is a direct reflection of their ultimate fighting method – raking the other cat's tummy with the back claws. Battlescars in old, intact toms tend to array the face and torn ears. The cheeks are very vulnerable to deep incisions from hooked claws and bad infections can result. Recovery from fighting can sometimes be a more dangerous business than fighting which has some consequence in the development of other signals before fighting is resorted to. Similarly potential prey can be dangerous to the cat. One cat of my acquaintance had a very swollen cheek for some weeks before a hedgehog spine was found in the wound.

Some indication of road casualties can be gauged from the 1968–71 figures of Baltimore's Municipal Animal Shelter (USA) for dead animals collected from the streets. The Municipal shelter is the primary agency for this work and has a twenty-four hour van service so employed. About 5000 cats a year were found dead on the roads of Baltimore, which was approximately 40 per cent of the combined dead cats and dogs found on the roads. This was in a city with a human population of about 900,000 people in seventy-two square miles of land. One cat found dead on the roads per year for every 188 people in that city. On Baltimore's rate of housing usage that is one cat found dead on the roads per year for every sixty dwellings (however an unknown percentage of these cats would have been feral).

Colin Howes found in Britain that spring/early summer produced most cat road fatalities, and thought that this might be due to more 'social' wanderings. Strikingly 70 per cent of the victims were black or black-and-white. There is a high proportion of black-and-white cats in urban populations, but are dark cats also more vulnerable to traffic accidents?

Although motor vehicles are a major cause of injuries received by cats, sometimes the inflicting weapon is more direct. In Sydney, Australia all radiographs of urban domestic cats taken at the University Veterinary Hospital and Clinic for a two year period were examined for evidence of firearm projectiles. In 180 cases three cats, all adult, were found to have been hit (any fatalities did not reach the clinic). More dogs were examined, but despite this, of the dogs and cats seen in both cases 1.7. per cent of the animals had been injured by urban gunshot. As in almost all of the cases (both dog and cat) the owners were unaware their animals had been shot, the difference in numbers would seem to be the usual case of cats being taken to visit a vet less often than dogs.

Most of the urban gunshot wounds came from .177. calibre air rifles, including those in two of the cats. It is the sort of weapon that children can often be found 'playing around' with. The third cat unfortunately had a leg amputated as a result of gangrene resulting from a .22 calibre bullet. It was believed that the malicious intent had been to kill the cat by this extremely dangerous weapon.

CHAPTER 13

Sex and the Single Cat

This chapter is aptly titled, for the historical view of the cat's sex life is essentially promiscuous. When mating occurs it is usually undertaken with a number of mountings and in consequence if more than one male is about the tom may change.

The young cat encounters puberty somewhere between five and eighteen months and in consequence I have come across some very small pregnant cats leading a feral existence.

In women between puberty and menopause there is a more or less regular cycle of twenty-eight days at which an ovum is released. The ovum journeys to the cavity of the uterus where it will remain for about fourteen days, anticipating fertilisation by a sperm.

In the cat the cycle is exactly half the time, about fourteen days and occurs in a slightly different manner which is reflected in both structure and behaviour. The queen does not ovulate in her cycle unless she is stimulated by the action of mating.

The female cat undergoes emotional changes during pro-oestrus and oestrus, the first stage lasting a couple of days and the next up to half a dozen days. People living with intact domestic queens will notice an increase in affectionate behaviour which gradually becomes more intense with rolling actions on the floor and calling cries.

The same type of behaviour can be seen in both feral and house cats, with an increase in chinning behaviour occurring with ecstatic head twisting and rolling about a marked spot. The throat is stretched out almost flat to the ground with the chin rubbing on the marked spot.

Strangely enough some neutered queens still occasionally demonstrate such behaviour but not usually to the extent of calling. (However varying proportions of the reproductive tract may have been removed, leaving residual hormonal function.) Our neutered female has on a number of occasions entered into ecstatic head rolling on a spot that I have been standing on for some ten minutes or so. At such times she has also attracted three or four males to loiter near her for some days. As she did not call during these times it is most likely that the toms were attracted by scent.

The initial attractant in free-roaming cats is unlikely to be from localised marking requiring close sniffing, for usually in the domestic ghettos of suburbia most of the males in courting will be trespassing in another male's range. However once in the female's range these scent posts will certainly be functional. It is likely that there are changes in the anal scent glands of the female which seem to have more volatile components at such times (detectable even by some people walking nearby). The action of anal scent glands in the cat are something of a mystery and it may be they leave a trace on the ground every time the cat sits with tail trailing.

As the males do not arrive together but rather build up in numbers gradually, and fights occur with loud intimidatory calls (especially pre-fight) these may in themselves be an attraction to other nearby males.

The attracted small group of males in the suburban house-cat pattern is likely to include the area male, plus adjacent area males, plus also the occasional stray. The area male will be more confident in his own marked range.

When the queen is sexually quiescent (anoestrous) she will not allow sexual approaches from a tom. When she passes first into pro-oestrous and then into oestrous she will progressively allow advances, but even up to mating she may make aggressive throat rumblings and sudden swipes at him with her paw.

Mating does not depend absolutely on the outcome of fights between the males, for the queen may have her preferences. Indeed during early pro-oestrous the first male on the scene may well make appeasing overtures to her. He may try to get closer to her in apparently easy little moves, only to freeze to the spot and play-act total disinterest the second she turns to look at him. It has a great deal in common with the childhood game where the object is to creep up without being seen to move. The close of the game differs however, for when the female cat considers that the tom has closed in her personal distance far enough she may pat him on the head with her paw. He will take this as he is able to go no further without her becoming aggressive or running, and·so sits, sometimes for hours at this conceded close distance. It seems that this courting can sometimes be the way to the lady's heart to become the favourite. Certainly males have been shown to have preferences in their choice of queen, between animals fully on heat.

As pro-oestrous proceeds the queen becomes more demonstrative rubbing against nearby objects, particularly with head and neck, and opens and closes her paws (as she does when settling down on one's lap). Males show progressive interest in her and attempt to mount her. These will be fiercely refused until her ovary is ready to ovulate. She will then enter oestrous and go properly into the lordosis position with rump held high, tail held erect with vigorous flicks and her chest and head held close to the ground. (This position again is a similar position frequently seen in a lap cat when you run your hand along its back). This is the position the male has been waiting for and he makes overtures to the on-heat queen by short chirp-like calls. If this is responded to the male will move in and take hold of the loose skin of the back of her neck in his mouth. This is done in the same way that she will carry her kittens around and in part of the same purpose to immobilise her. (Using the same technique a vet restricts the movements of a cat under examination.) The mount itself occurs in a very similar way to a horseman getting into the saddle inasmuch as the forepaws first find a hugging grip by her sides and then the male brings his hindlegs round to cover her body. As the front of the queen drops to the floor to further elevate her pelvis, he lifts first one back paw and then the other and as he steps he sways from side to side and rubs her sides with his front paws at the same time. During this he will increasingly arch his back until his penis makes contact with the female's genital area. She joins in with his treading motion with her back legs. The male then undertakes a series of vigorous thrusts of his pelvis and copulates quickly, in about ten seconds. Certainly this is the length of time under laboratory and stud observation, but in the free state a solitary male and female uninfluenced may continue considerably longer.

The male immediately leaps well clear for the queen cries out and strikes out with her claws.

Both male and female wash their urinogenital area, the female vigorously as she rolls around. However half an hour or so later she will allow him or another to mount her again and this will be repeated time and time again.

If only one tom is present, after the first few matings, rather than allowing the tom to mate she positively encourages him with seductive movements and as matings continue this change of interests increases, the male's interest in her waning and her interest in him growing.

Why the female screams has been a matter of debate for centuries with such colourful suggestions as the tom having 'seed of fire'. However investigations before the turn of this century as to a possible cause produced a possible answer no less dramatic. Attention was drawn to relatively large pointed horny spines on the penis of the adult male cat. These spines when examined under scanning electron microscopy seem to have very similar appearance to the papillae on the tongue of the cat.

These penile spines increase in size as the male hormone levels rise and decrease in size as the male hormone level falls through the animal's life. In 1934 under research conditions female cats were artificially stimulated and reasonably strong evidence was found that they ovulated following mating (about twenty-five hours later). So it seems that the raking by the large spines of a tom in his prime probably increases the likelihood that she will ovulate while such a fertile tom is about. Her attracting many suitors, combined with the tom's fighting to gain her attentions ensures the strongest possible kittens. Her early delaying coquettishness ensures the largest number of males present and therefore the likelihood of toms in their prime to be there when she is ready to ovulate. Despite all of this her body needs time to deliver the eggs to the uterine site to be fertilised, so her apparent unsatiated behaviour allows time for the physiological events and is not as some would suggest evidence of depravity.

The distortion of the spines during intromission has been suggested as stimulating the male and sexual interest and behaviour was found to relate to male hormone level in cats and thus to the size of penile spines. Therefore the tom in his prime as well as stimulating ovulation in the female is also highly likely still to be present due to increased sexual interest, more so than juveniles or aging toms, up to a day or so later to fertilise her.

Strangely enough debate continues, does she scream when the tom enters or when he leaves her, and the final barbed question – is it pain or could it be pleasure?

Where to operate

When undertaking surgical neutering (ovariohysterectomy) of a female cat the route by which a veterinary surgeon cuts is strangely enough more likely to depend on whether he is British, American or Australian than on more fundamental surgical rationalisation. From his training he may think a mid-line incision on the abdomen to be best, for it is certainly the easiest route by which to find the uterine horns (with the added advantage that many cats are nearly bald at this point). Alternatively he may have been taught that cutting in through the flank is preferable as hernias are less likely.

Where the operation is to be carried out on a domestic cat is best left to the vet and the way he has perfected, rather than try to force an unfamiliar practice onto him. However, when dealing with feral cats the situation is slightly different. I feel the incision on the underside to be the one then of choice, as there it is virtually invisible. A shaved flank on the other hand excites notice among the human population. Such is the variation in human emotion that an apparently injured or diseased cat (as the shaved area of flank can give the appearance to the undiscerning at a distance) could elicit anything from a lobbed half-brick to attempts to corner it 'for its own good'. The hair takes a few weeks to grow back properly. We have had both types of operation undertaken on females returned to a colony, midline on one half, and on the flank of the other. In the identical situation on their return neither half fared worse than the others, but nonetheless I feel that it does constitute a real danger and is easy to avoid.

It is particularly essential to use soluble sutures, such as 'catgut', if at all possible on feral cats to avoid their needless disturbance by recapture. Alternatives to the full ovariohysterectomy is the use of ligature on the tubes or removal of a section of the tubes, leaving more hormonally intact cats.

Left to their own devices, suburban domestic intact queens will have two (or possibly three) litters a year, usually in the spring and autumn.

Free matings can result in kittens being born during the winter months, although to a lesser extent than in the rest of the year. This is supported for feral colonies in England by the professional cat catcher, Mike Jackson, who has found dead kittens in nests, born during the winter, that were unable to survive.

Combined with such seasonal young deaths the sexual activities of both male and female suffer an understandable decline during the winter months giving a distinct seasonal cycle to successful rearing. Latitude and day length seem to be likely controlling features.

The walls of the vagina and uterus undergo a regular cycle of preparedness for the implantation of the fertilised ova. Following implantation this cycle is interrupted to allow development of essential communication between mother and foetuses. However, in a sterile mating by an infertile tom not only is the ovary stimulated to release the ovum, but the walls of the uterus undergo for a period of 20–30 days the events as if the cat was pregnant. This pseudo-pregnancy shows again the powerful effect copulation has on the female cat. It also may account for a number of apparently pregnant feral cats seen during the winter, in which there is no trace of kittens being born later.

Not just the voice of experience!

The male sex hormone originating from the testes (testosterone) and experience in mating are very important. Normally at the magical age of 333 days the average male cat can start mating. What happens to the sex life of the neutered male as the circulating testosterone declines following surgical removal of the testes? The unfortunate adage 'What you've never had you don't miss' would seem to have some foundation in this instance. Two distinct experimental methods in an American study revealed this. In one, animals with and without experience were neutered, while in the other method, animals were neutered while still young kittens, but by then giving testosterone artificially were able to sexually mature, with one group being allowed experience, the other not. In both methods as the male hormone level dropped it was the sexually experienced animals that continued longer to behave sexually and mate. Considerable variations are known to exist between different animals after castration or hormone withdrawal, with some showing no loss of intromissive ability for as long as four and a half years, while others cease to pay attention to females after a few weeks. The latter type of neutered tom does not seem to suffer frustrations at being deprived for he tends to lose all interest in female cats inasmuch as he even stops approaching them.

Discussion with many owners of neutered toms has revealed that the normal neutered domestic tom leads a fairly happy life, perhaps a bit playful, but by no means hounded to distraction. On occasions however one meets a lady who says that her particular cat has suffered persecution from intact toms because he is neutered. It is strange how in 'cat gossip' circles it is this latter opinion that is so often voiced as a generality.

It is worth noting that neutering can affect confidence. What is often overlooked is that not only can experience before castration change the cat's sexual performance after the operation, but it can also modify less obviously sexual behaviour. For example, a tom neutered without sexual experience may be frightened when confronted by a queen on heat. In contrast a sexually experienced neutered tom may well respond to an oestrous queen by playfully rolling with her and patting her. He can seem like a kittenish flirt. Prolonged mounting seen in post neutered males usually derives from insufficient erection rather than from full vigour.

During the period that mountings still continue in the post-neutered male, if intromission does not occur, as is quite frequently the case, the female may on occasions turn on the male to which the male may retaliate. Alternatively the neutered male may just roll off following a prolonged mount with no intromission. About a quarter of neutered males still achieve full intromission at three months after the operation.

Cats and kittens

Cats tend to have on average fewer young than similar carnivores, including dogs, and when the kittens are born they seem to be at a slightly more developed state than puppies. The litter size may relate again to the social structure of the cats. The male is unlikely to act as provider of food to the nursing mother and kittens. Indeed due to the number of different males that may have mated with her when she was on heat that derived from the requirement for ovulation to be sexually stimulated, it is hardly surprising if a flood of paternal feeling does not swamp most toms. In the absence of social provisioning the female has to suckle, feed, teach and protect her young normally by herself. Smaller litter sizes of around four kittens then would seem to have evolved with the behaviour pattern.

The mother's responsibility lasts at least the nine weeks or so until weaning, and in colony cats the close association usually lasts longer. However, up until the time for forays the kittens will stay in or around a den. Although cats do dig holes for faeces, they have not developed into burrowers and tunnel excavation is beyond them. Nonetheless a snug small cave is a preferred nesting site.

In its absence the urban feral cat finds many ideal structures around man's buildings. Wiring culverts under hospitals are ideal cave replacements and under factory storage bins the eight inch or so gap at the bottom that is wired around to keep rats and cats out is usually broken in one or two places giving a much safer den. Prefabricated temporary office buildings provide excellent den quarters, as netting is again usually put all round the larger gap under the building and similarly there are always a couple of holes. I have also seen dens under the long hollow seats of factory locker rooms.

Garden sheds standing a couple of bricks high off the ground, brick piles and timber stacks have been used by nursing mothers. The nearest to a cave I have seen used was an old bin that was partially covered up with soil lying amongst a spinney of trees. The mother may feel the need, if disturbed, to move her young, and possibly the most potentially transitory den I have come across was inside the wheel arch of a Mini-car in a garage!

Kittens stimulate the mother's milk flow by alternating kneading movements of the front feet, the same movements that are seen in the contented animal later in life. The young of many other mammal species also employ this early aid. It is often thought that domestication involves a gradual selection for animals retaining juvenile traits. The argument is largely based on the belief that young animals are more docile.

The kittens, again in common with some other mammals, purr while suckling at the teat. As a reassurance sound it is also made by the mother cat to her young.

The mother domestic cat initially starts the feeding sessions while the kittens are very young, during the first three weeks, but by five weeks the demanding youngsters have turned the tables and from then on they are increasingly making demands on the mother.

Even so, the kittens come to 'own a teat' within days of birth which is just as well for even at that age their claws could easily rip a teat.

Domestic Cats – progress from birth to 6 months

9–20 days	*32 days*	*8 weeks*	*6 months*
Eyes open	*Eating solids*	*Weaned*	*Independent*

One product of domestication seems to be an increase in number of litters per year, for most of the cat family only give birth to one brood per annum.

CHAPTER 14
Are There Too Many Cats?

To estimate the number of feral cats for as large an area as England or the USA is very hard and with any degree of accuracy virtually impossible at present.

It is not even easy to determine the domestic cat population. Pet food manufacturers have been looked upon as a reliable source for this figure, based on their sales and average consumption figures. However, due to the huge army of cat feeding ladies that in large part maintain urban feral colonies the distinction is not clear cut, as many of these ladies buy tinned cat food.

Nonetheless using the survey findings of cat ownership (Chapter 5) and census figures it is calculated that the UK has a population of domestic house-cats of about three and a half million while Pedigree Petfoods estimate it to be nearly five million. This is only between six and eight cats per one hundred people.

Despite this it is sometimes suggested that because of the combined amount of food (and its cost) that pets consume they should be reduced in numbers, for they are eating and third world children are not. This is a dishonest argument for the status quo continues on a larger scale as a reflection of personalised greed.

In the UK in 1979 prepared petfood sales were only 4 per cent of all grocery sales. In direct terms pets eat a small amount compared to the great excess of food consumed over need by western man. If there is any doubt as to this only a cursory glance need be given to the statistics for diet-associated diseases to which western man succumbs.

The urban stray and feral cat by its scavenging activities is helping to reduce part of man's edible waste. Nonetheless, most people agree that there are too many such free-living cats about and much money and time, generally uncoordinated, is spent trying to reduce the loose cat numbers.

Is this action justified? Currently the guess for the feral cat population of Great Britain is about one million. I would put the figure higher and using available figures suggest a population of urban feral cats and suburban strays for London alone at about half a million. This is far from definitive but will have to stand until more information is available. This is also true for my estimate of the UK feral cat population of at least one and a half million cats. One problem when considering the rural component falls on an old chestnut of a 'matter of definition' and that is with the semi-feral farm cats: when feral and when not!

It is certainly not even easy for the unaccustomed eye to make a reliable estimate of the size of the cat population over even a relatively small area, such as a hospital grounds, even when the person involved works there. There too seems to be often a perverse need to elaborate in the human character so that a real population of twenty cats may be seen as sixty and forty as 'hundreds'.

A number of both commercial and animal welfare cat trapping agencies exist and these can provide numbers, but only of cats caught. Fortunately it is possible to make a reasonable estimate from such figures. For example, 1610 feral cats were actually caught from fifty-one factory and hospital sites in England, but by applying the Standard Minimum Method I estimate that the total population was 2234. Overall 72 per cent of the estimated populations were trapped. When each site trapping efficiency was found individually, then the mean efficiency overall was 81 per cent.

At these fifty-one hospital and factory sites the mean number of cats per site was estimated at forty-four, although the mean number caught was thirty-two.

When the mean numbers trapped on different locations are looked at they give some idea of relative colony size. Rural hospitals were found to have a little larger colony size than urban sites. On urban sites factory colonies (mean sizes) were smaller than hospitals. Residential sites had very similar mean colony size to factories, but commercial properties (shops, etc.) could not quite support as large a group.

The feral cat is not necessarily a health risk to either ourselves or our pet cats, indeed they try to keep themselves clean and healthy and away from us. Nonetheless certain colonies and certain cats are a risk. People who make it their business to catch these cats are obviously exposed to a greater risk. One lady I know of who regularly both traps and feeds in one area of London has had ringworm three times since she started. In the same way cat colonies that are established in locations where health risks must be minimised, such as in hospitals, warrant some control.

In terms of colonies, even if large, that are isolated and not causing any harm, such as the one studied by Jane Dards at Portsmouth dockyard, then as she suggests they may as well be left alone. The colony number will stabilise between certain levels depending on such factors as availability of food.

In many urban settings however the optimum number is usually higher than the local residents wish to tolerate. In many cases however the objectionable feature is the spray of intact toms in stairwells of blocks of flats or against doors. This is not necessarily a feature of high numbers and could be dealt with as a single entity.

Certainly undesirable effects accompany large numbers of cats. Mess in one form or another and smell are among the most frequent reasons given to professional cat trappers (pest control) companies for calling on their services. (A further reason often grouped with these, is the noise of catawalling). The mess does not always require high numbers before industry calls on cat trappers. One East London furniture manufacturer was suffering hundreds of pounds worth of damage to finished furniture from cats' urine. Following a trapping there were found to be only five strays on site.

The emotive issue of trapping cats brings an interesting slant to industrial relations. In many instances trapping is undertaken following strong requests from the men through their union representatives. This can be due to the working conditions for the men being affected to the extent of hazarding safety. In one immaculately clean car manufacturing factory, due to the lack of any sawdust or similar material that the cats could have employed as a latrine, a fork-lift truck skidded on a pile of cats' faeces resulting in injury to an employee. In contrast in some instances where management takes an unnecessarily high-handed attitude and instigates the trapping and killing of all cats on site understandably some resentment is voiced by workers. For on every site there are often one or two special 'pets' among the cat population and these are effectively domestic animals. More enlightened companies first of all have combined union and management talks to establish the complaints and to determine which cats, if any, are to be left on site. Such cats can be fitted with a cat collar, and if the cat is caught it can thus be recognised and released. However such 'company' cats should be neutered so that a recurrence of the problem does not happen faster than need be. Indeed, one of the large car manufacturers has introduced the policy of recognising certain cats on site officially in all its factories in Britain and these must wear collars, be neutered and have properly constructed kennels provided.

The British Museum has responsibly pioneered this approach by instigating a committee of interested people on the staff, then by declaring certain feeding sites official, reducing random overfeeding, and providing large comfortable waterproof kennels on the flat roof of a single storey building. There the neutered cats are undisturbed when sleeping, and consequently make full use of the cat-houses.

Indirectly a combination of faeces and flea infestation is often put as a health hazard in requests for clearance, such as by hospitals. Even after cats are trapped their fleas can cause a problem, as at Covent Garden, where building work could not proceed until the fleas were eradicated.

Hospitals may present a specific case for cat number control and indeed with the requirements of the Health and Safety at Work Act both sides of industry now feel the responsibility.

However, in other locations that are more open is there a need for present controls with the spectre of rabies looming from Europe? As the feral cat will be so hard to control if rabies were introduced it makes strong sense to reduce numbers now to manageable proportions. However, it should not be forgotten that most animals can carry rabies (including mice!) and it would be rash to eradicate our entire wildlife on that basis. It also prompts the question at what point could a feral animal be of such numbers and importance that it could claim to be part of our wildlife. Few people would doubt that the feral pigeon has reached this status and I would suggest the feral cat has this as well.

Nonetheless, if the arguments hold that stray and feral cat numbers should be checked, what methods are open to use?

Catching cats

Eradication from a site is an obvious approach. Nonetheless there are many ways. At least one pest control company shoots cats with a rifle in rural settings, but such methods would not be safe in towns. For a while another pest control firm shot cats with bolts from crossbows, but a court case resulted in this being abolished. Shooting is obviously not humane when a clean kill does not result. Poisoning with a narcotic such as alphachlorolose, then painlessly killing the cat subsequently has been suggested, but it would be unusual to be sure that the cat would not wander off before narcosis was effective. Fortunately legal restrictions exist on poisoning.

The almost universal method employed by pest control companies and animal welfare agencies is live-trapping, whereby the animal enters a cage searching for bait and is restrained. This can then be followed by painless destruction of the animal, but does not have to be, for the live trapping method is the basis of many of the alternative courses of action as well.

Most trapping is undertaken at night, which generally seems to be the preferred activity time of the animals with regard to feeding. The main activity times over the year are 9.30–12.30p.m. and 3.30–5.30a.m. but particularly the evening.

If the operator is certain the trapped animal is feral a decision may be made to kill the cat. It is usual to put the animal into a chloroform chamber where over a period of five or so minutes the cat is overcome by the vapours. Having witnessed this procedure many times few cats seem to fight the vapour, appearing to fall into a drugged sleep, but some large cats do not willingly bow to narcosis. A carbon dioxide euthanasia chamber has been put forward as an alternative, but at present this does not seem much more humane. Sometimes the caged cat is taken to a nearby vet for destruction by injection.

Eradication from a site also includes removing the animals for attempts at housing or to a sanctuary. There are many advocates in the cat welfare world who are strongly in favour of sanctuaries and homing regardless of how wild the animal seems to be. Within the same world there are equally severe critics of this policy who say that the huge number of deserving cases of recently abandoned cats, many already in animal welfare catteries, that both need and would appreciate housing make it heartless to try and home wilder cats that have already shown their preference for avoiding human association in their stead. The vast number of totally free living cats certainly make the latter a more realistic proposition, but the people in the former camp have a missionary zeal in their outlook. The suggestion that free-living animals, perhaps into generations of such life, especially in rural settings where the territorial range is so large, might not really appreciate being 'saved' by being put into the narrow confines of a penned cattery is a strong bone of contention. That animals can be tamed to consort with man is not a valid argument for their being happier in that state.

In a rural setting eradication by whatever means can effectively control numbers for a long period, as can also happen in isolated urban sites. However in most sites in London where colonies have developed, if the site is good enough to support a healthy colony, eradication merely leaves an effective vacuum that envious cats around the corner are only too keen to fill. Therefore eradication policy working slowly and patchily through a huge population over a huge area such as London is doomed to failure. This is effectively the policy for London that has been executed by tacit agreement by many animal welfare organisations and individual trapping ladies and pest control firms for the past twenty years or so. The parallel to painting the Forth Bridge begs to be drawn.

Population recovery

While trapping and subsequent killing of feral cats is an emotive subject, hospital feral cats give the issues an added twist to the tale.

Currently most hospital administrators take the clear-cut view that such cats are carriers of pests and disease and total site eradication is the only course open to them in the interests of patients' health. Certainly reported cases of cat-flea infestation of surgical wards are disturbing.

Nonetheless in most mental hospitals, which usually have fairly extensive surrounding grassland, the presence of a cat population is not only tolerated but often encouraged – up to a point – as being therapeutic to the longer term patients.

A growing number of administrators are following a policy of neutering and partial return of the cat population, particularly of the more amenable healthy individuals.

From the cat's viewpoint hospitals are almost ideal sites, as they invariably consist of rambling buildings with culverts for wiring and pipes ramifying underneath large areas of grass, shrubbery and earth.

However, how effective is trapping in keeping a hospital cat population of zero? It certainly is not as 'clear-cut' as many administrators would like to think, for they do not 'get rid of the cats'.

Sister Sheila Young maintained a group of 3 neutered cats at the Middlesex Hospital for nine stable years. A new administrator insisted on their removal, after which the 'vacuum effect' operated and within eighteen months sixty cats were living in the hospital grounds. Sense then prevailed, and Sister Young again maintained a small neutered group.

For hospitals for which figures are available of an initial trapping followed by a subsequent trapping 5–10 months later, the cat site population was recovering to a greater or lesser extent about two-thirds of their original number.

At some hospitals, on finding the population building up again after extensive trapping both from those remaining and invading cats, they have resorted to using the pill at 'official feeding sites' to avoid this.

Toms in first!

In a sample of ninety-four adult feral cats from colonies, examined specifically for their sexual status I found near identical proportions of 52 per cent female and 48 per cent male overall. However, it seems that colonies that had not been disturbed for a few years by removal trapping had around 35 per cent adult males

(20–50 per cent) while in colonies where recent removal trapping had occurred the males were in the region of 60–75 per cent.

Taking as an example an industrial site colony in Leyton, north-east London, on complete trapping-out and removal it was found to have 50 per cent males (twenty-two cats). Twenty one months later when the site was re-trapped out the fresh animals that had recolonised the same location had 75 per cent males (twelve cats).

This apparent initial predominance of males on recolonisation would seem to go along with our new understanding of the home ranges of the sexes where the bold far-ranging male contrasts with the more tightly home-based female. Males are in a better position to act as the first re-colonisers of an empty site.

Effective control

The best hope for effectively reducing overall feral numbers lies in a combination of actions. The first is to avoid the topping up of the wild population by dumping. Topping-up of free-living populations by dumping of kittens after weaning in hospital grounds, factories and similar places is unfortunately all too prevalent. Indeed in colonies that I have been watching, this can be of the order of 10–20 per cent of the colony size per year. However in long established colonies these are unlikely to be accepted for months, if at all, but in loose structured recent groupings of cats they are quickly assimilated.

It is clear from the Romford Survey (Chapter 5) that most cat owners are responsible and neuter their pet. As a large number of cats were seen to have been taken in from the fully free-living population, then this in itself is one of the major controlling factors. If a greater proportion of that sector that do not neuter could be encouraged too, much of the feral 'problem' would fade away.

This depends on changing people's attitudes. Many owners of intact toms (that are not required for show breeding) are socially irresponsible. Their exploits and spraying can cause anguish for neighbours. The keeping of intact queens perhaps is based on some owners insisting on at least one litter. This is perhaps to prove the queen's sexual capability – that she is 'all right'. Keeping a queen intact and then not allowing her to mate is far more dangerous than neutering (which can extend the cat's life) for in such a neurotic state some female cats develop pyometra (seen as acute uterine enlargement, accompanied by abdominal swelling). Changing social conditions may change these attitudes, but having a cat licence, at least for intact animals, would certainly help. If the licence fee for having an intact tom or queen were much higher than the normal veterinary fee for neutering, then much of the problem of unwanted kittens would disappear. People who enjoy breeding cats and have a responsible attitude to cats would not mind paying for the privilege. My one big fear of this approach would be that this would result in the 'moggy' varieties of cats reducing in the home relative to the less genetically healthy 'fancy' varieties.

Breeders of 'fancy' varieties may wail, but the official Swedish cat bodies' fears about the spina bifida like presentation in some manx cats is not misplaced. Breeders may practise selection by killing or not breeding from animals with obvious genetic defects, but selection is nonetheless practised within an inbred stock. Continually selecting for a limited number of features is a dangerous game, and can defeat its own end. Some of the beautiful white long-haired cats that are bred for that endearing snub-nose look have now become a pathetic spectacle as the secretion from the eyes wells down resultant infected and matted cheeks. A certain amount of random mating is much safer.

Another arm in reducing overall cat numbers derives from our growing knowledge of cat colony behaviour. Remove totally a cat colony in London and over a period of months that territorial area will have been refilled by a new set of cats. The conditions that enabled one set of animals to live will adequately tolerate another group. Therefore the cat vacuum of clearance should be avoided by allowing a number of cats to remain but that are not breeding. The two methods of achieving this are putting puss on the pill or neutering.

Cats and contraceptives

The oral contraceptive pill has been tried in field trials on feral cats in Britain. In a recently abandoned hospital in Hertfordshire where fourteen cats continued to live (of which seven were females) the progestagen pill was introduced into food offered to the animals at feeding points. After eighteen months the trial was concluded as the vagaries of movements of the cats made the possibility of administering a regular and accurate dose unlikely. A further complication arose of one of the animals developing pyometra. Nonetheless it is being tried on some populations of other hospital and industrial sites in Britain.

In Israel the pill has been given in the same way, once a week, to stray cats for fifteen years. (The dosage was established by the Hebrew University Pharmacology Department, the drug is distributed by the Israel Cat Lovers Society). Currently it is being given to about 2000 cats (within a 'guestimated' population for the country at least a hundred times larger). The society believes they are preventing the birth of some 20,000 kittens per year. Nonetheless 5 per cent of the cats have been seen to have endometritis. Effectively this condition can be thought of as a less acute form of pyometra. It is usually recognised by vaginal discharge and consequent repeated licking of the vulva.

The unfortunate problem with both of these studies is that the pyometra and the endometritis could have occurred anyway, or they could have been caused by the pill.

As a system of population control it is only possible with regular feeders who can ensure that each cat obtains its separate dose. Weighed against this is that it can cause reversible sterility and this has proved attractive to some Danish farmers for their farm cats. It is perhaps in such controllable situations that the future of pills requiring regular dosage has its main application. The development of a 'once and for all' pill, very carefully administered, might remove worries of any variance in feeding patterns.

The unkindest cut of all?

One major advantage of neutering by surgery is that once it is done it is done. It is not dependent on playing a form of Russian roulette with dosage.

The animals are trapped in live-traps and immediately covered with a blanket, which quickly calms them down. As trapping is invariably carried out at night a short term hostel is required for the cats until the following morning when they can be examined by the vet. However, as they may be carrying disease this must be an isolated pen which can be thoroughly sterilised after their departure. On examination any animals that are very ill and suffering can be put down, but in healthy animals the operation can proceed.

It has been found from long practise with domestic cats that they tolerate this operation very well, regardless of their age (over 7–8 weeks). As yet no system of marking neutered cats has been formalised in Britain such that some cats previously neutered risk the possibility of being re-opened, certainly of the unnecessary trauma of trapping and transport. This occurs with house cats as well that they have strayed in their life from other owners, so a universal marking system would have considerable advantages. A coded tattoo could convey far more information about a cat's origin and treatment than a single mark, but the obvious place of the skin inside the ear is often so pigmented as to make this unclear without close examination. Especially for strays it is essential a fairly clear mark or attached tag is visible so that cats caught on subsequent occasions by mistake can be immediately released rather than wait until the animal is anaesthetised to find out. On the other hand it should not disfigure the cat or set up irritation.

The Danish Cat Protection Society have developed a code tattoo on the inside of the ear, which according to vet Tom Kristensen had been used by 1980 on 20,000 house-cats with minimal occurrence of infection. The code identifies the vet, the area and the individual cat. To be able to tell at a distance that a feral cat had been neutered (and also vaccinated for distemper and 'flu), it receives in addition to the code, the tip of the other ear clipped off. To some this may sound a bit drastic, but at least it has shown itself to be clearly recognisable. The Danish Veterinarians' marking accompanies the operation.

A programme of neutering and partial colony return has been taken up by the Cat Action Trust in Britain and seems to be being successful. I had the honour of launching the C.A.T. but in my address voiced reservations as to how well a group would stay together once neutered. Cat territoriality is scent marked and scent production is under sex hormone control. The study in Fitzroy Square (Appendix) was largely to see the effect on the cohesiveness of a neutered colony. As it has turned out the colony stayed together, and continued to hold its territory.

Totally neutering all the feral cats at a site may in the long term be almost as bad as killing them all. It could mean delaying the time until the cats naturally die, and new strays enter to fill the vacuum. If the colony is a long term established family group it would be better to keep a few intact to perpetuate that close family line with its social bonds longer than one generation. Most successful biological control depends upon a population continuing at a low level rather than total eradication. However this will demand a very careful watch being kept of sites, keeping the population low, and neutering most of any new kittens. Here is perhaps a new demanding role for the concerned feeding ladies.

Even with a policy of neutering both feral and domestic populations, one factor alone could make or mar any programme. Although cats are superb survivors and scavenge as well as hunt, it is mainly the massive auxiliary feeding by kind-hearted people of feral cats that reduces the need for large home ranges and allows the far higher densities of urban feral cats above the rural population level.

Unfortunately by excessive feeding dependence does build up. Not only will a group increase, but being fed, its home ranges decrease and gaps are left for more cats to increase the area density. Sudden stopping would certainly cause hunger and distress when numbers have built up.

The ladies are acting for what they see as the best needs of the colony and so will often co-operate with a neutering and partial return policy of animal welfare organisations. As long as they do not overfeed but gradually reduce their feeding amounts over the years with reduction of colony size, the cat numbers should gradually and painlessly stay smaller. The cat feeding ladies hold the key.

If a site has been cleared then feeding should stop, but if it has had partial return of neuters feeding should be reduced. In either case it is essential that the reasons are clearly given in the local press and perhaps a local notice, so that the whole thing does not repeat itself by someone else 'helping the cats'.

Summarising, the feral cat population can be diminished when required by neutering and reducing colony size, and reducing extra feeding, plus increasing the number of neuters among the domestic house-cats.

However the cat is a survivor and will strike a level with the available food.

'Dust to dust, ashes to ashes. . . .'

Frequently more hypocrisy accompanies the practical arrangements that follow the death of a pet than follows even the death of a relative, if that is possible. Nonetheless in both cases I am not alone in feeling that it is justified, for although looked at disinterestedly by a third party the prime requirement may be a public health need for rapid disposal of the body, the emotions of the bereaved must be allowed to be expressed.

If a cat has been a true part of your household for many years, then it is natural to feel the loss of one of your family unit. As with the death of a person some people tend to dwell more on the past than on the future. One lady I know of has been visiting each week the gravestone of her pet cat in a north-east London Pets Cemetery for sixteen years, and she is probably not alone in such prolonged devotion. However protracted her attachment to her pet is, she has taken the responsibility of personally providing for its place of rest (or for its hygenic disposal, depending on your involvement).

Taking personal responsibility matters if you wish to duck the hypocrisy. The person who says to the vet in a subdued voice 'Will you please take care of the arrangements for puss?' after puss has been 'put to sleep' is understandably not looking too closely at the details.

If the vet buried every cat/dog/budgie/gerbil in his back garden he would long ago have run out of soil. Individual incinerators of sufficient size to burn a cat would cost a fortune, be very large and in residential neighbourhoods would not be favourably received. What can he do?

In some areas, but by no means all, local authorities will collect animal carcases as part of trade refuse (and then often at a charge). If the local authority operates a large incinerator for refuse the animals will be thrown in. If not then where controlled tipping is occurring for land reclamation the local authority will tip the bodies, along with condemned food and other such material in front of the tip face and cover immediately with other refuse, so that the bodies are at least three feet under the surface. In certain areas such as east London, specialist companies accept animal corpses from vets, animal welfare organisations, pest control firms and similar people who by the nature of their activities end up with animal corpses to dispose of. Those disposing do not gain financially, indeed quite the reverse in terms of petrol and time on a regular basis. Then the corpses are turned into such animal products as hoof and horn, bonemeal and similar fertilisers.

For an animal that one has not been fondly associated with this would seem to be a reasonable end, and not really removed from 'dust to dust, ashes to ashes. . . .', as the fertiliser will be scattered back on the land. In the same way the local authority is really burying or cremating, it is just that it is termed tipping and incinerating. On public health grounds this is certainly fine for strays or for pets whose householders do not mind. However for some it is not just a difference in words, it is a difference in how it is done. For people who do mind, they must take responsibility for their pet's body – and this can mean financial responsibility. If you have a garden you have no problem, but if you are a flat dweller and try to find a plot of land in a pets cemetery you may have the response 'No room – full up'. The same is even true for burying people. Many of London's human cemeteries have been full up for so many years that due to consequent falling income the private companies that run some of them are demising as well. The problem has been sometimes circumvented in both cases by the installation of large incinerators, that is crematoria.

The present more usual charade at the vets puts a moral burden on to him, forcing him by collusion into the role of almost a liar. If an old lady with a cat that has been her sole companion for up to twenty years needs reassurance, then a few words sympathetically spoken at a cremation in a local pets cremation centre are a reasonable request. There does not *have* to be a theological content in any such brief remembrance. The ashes scattered in its garden will be more comforting than wondering what the vet really did to puss; she will be left with something tangible, a location. If this is the need then payment must be given, at least to cover the overheads, otherwise the charade will continue, but perhaps more realistically. It will also do away with a genuine public health worry in some areas.

Conclusion

The cat's domestication awaited man's domestication. The advent of reliable scavengable material with man's urbanisation may indeed have provided the selection pressures to 'invent' today's cat. The ready ability to accept scavengable material in the street predisposed the cat also to accept food on a plate in a house. The superabundance of food led to a higher density of cats (with smaller home ranges), undoubtedly proving that domestication has been a survival gain for the cat.

Nonetheless, the cat has not forfeited its independence. If it can exploit a food source it will do so. Although the fireside loving, but free-living house-cats, and the less cossetted farm-cats plus the rakishly independent feral cats of town and country, may seem to lead very different lives, when man is seen in perspective as a near surrogate cat then even the pattern of land group use remains constant for the different types of domestic cat.

The domestic cat despite its shades of domestication and ferality remains the supreme mammalian hunter. Yet just as its independence led man over the centuries to endow the cat with divine or demonic capabilities, so the cosy fireside image of the cat has blinded us to the fact that the cat is also an integral part of our wildlife.

We like the idea of 'wild' nature, and try to pretend that we do not live in an intensively man managed landscape. To many urban ecology is not real and feral animals somehow not there. That is to ignore a very real presence of Britain's predominant predator – the cat.

Few doubt that the grey squirrel and the rabbit are an integral part of our wildlife and yet both were introduced to Britain after the domestic cat. The nature of the cat means that we have had feral domestic cats almost as long. The feral pigeon scavenges in close association with man, yet few doubt they can survive or suppose that man's semi-supportive role is wrong.

The domestic cat is with us, the feral domestic cat is certainly part of our wildlife and has been so for a long time whether we admit it or not.

Even to recognise that however is only to acknowledge a half-truth, for the domestic cat in all its forms, is a major predator on other species. As such it is hardly a new impact, just filling the place of the declined wild forest cat – not a devastator of our wildlife – but part of it, and yet partly a house-living population as well. This is the messy end of ecology, where the line of impact between wild and domestic cannot be drawn. However, it is only messy because we construct an image of the natural world as if we were not part of it. The cat accommodates us by living in both worlds, even to the extent of fitting us into its group or colony. In the cat we can see man as just a symbiotic species, and domestication as just one ploy in its survival game that has worked.

APPENDIX

An Evening in Fitzroy Square, London W1

Although all cat colonies vary, they have enough points in common to make it worthwhile to sit through a fairly typical evening with one group to watch what happens.

What follows is a summary of one night taken from my many evenings observations over several years of a colony of cats living in Fitzroy Square, in the west end of London.

7p.m.	No cats visible in square.
7.30p.m.	Two have bounded into the bushes from a basement.
8p.m.	Still only those two cats.
8.30p.m.	Six cats in the square (only one in primary area).
9p.m.	Six in square, two in primary core.
9.22p.m.	**Cat feeder arrives.**
9.30p.m.	Six cats in primary core area, five at main feeding station with feeder.
10p.m.	Eight cats now in primary core area, only one remaining at main feeding station but one other eating at railings.
10.30p.m.	Only three cats remain in square and only one in primary core area.
11p.m.	Only one cat left sitting in square (in primary core).
11.30p.m.	All gone.

On these bare bones of the evening it is worth looking at bits of the flesh.

	1st June
7p.m.	Initially fine warm evening. No cats visible moving in square. (Only cats about are three in a road off the square belonging to an adjacent group whose range does not extend to the square. One of these scavenging (chicken bones) in a skip while other two sitting under a car.) One black and white cat asleep in his favourite spot on ducting outlet on building on corner of the square safely behind railings.
7.30p.m.	Two black and white cats ran from the cellars of a house (M), played, and went into bushes (A).
8.04p.m.	One out from bushes. Another cat of this group runs through the seats (H) then dives down through the railing into the basement (H), of what was once the home of a prime minister.
8.15p.m.	Mother (human) and her child playing football in paved section of cats' core area of the square (ten minutes).
8.25p.m.	One black and white cat at tree with a scavenged piece of fish, and another of the group digs a hole in primary core area between centre tree and railings and defaecates, then fills in hole with seventeen scratches on ground. Moves back. Another cat bounds into square.
8.28p.m.	Tabby/white cat moves cautiously into centre of green area. This cat not a group member, but has kept peripheral to the group for a couple of months, trying to gain access. Tabby/white moves off across path towards bushes (W).
8.30p.m.	Although fewer people about now and children gone, nonetheless about six people in the square at any one time (in groups of 2–3).
8.35p.m.	Cat defaecating in flower bed by bushes. Dog being taken for walk moves up to railings and cat retreats, to return when dog moves off. Another dog moves through square.
8.42p.m.	Cat runs out of one basement and into another.
8.45p.m.	Tramp sitting in seats (H) for half an hour, while two girls eat fish and chips from paper at seats (L). One cat moves back to basement railings from bushes (A), but then returns to bushes. Another cat defaecates in the earth bank by bushes (W).
8.50p.m.	Third dog of the evening taken for walk, defaecates in gutter (near the bushes at (A)). Cat runs from bushes to

pavement area (V) looking in the direction Mary (cat feeder) will approach from. The dog disturbs it and it returns to safety of grass behind the railings where after a few minutes it sniffs and eats some long grass. Following this it moves across the square to sit on the windowsill of the house where Virginia Woolf once lived (Q).
Two cats seated at Tree B.

9–9.15p.m. More cats come into the open, a couple sitting on pavement looking in the direction the feeder will appear. The little anxious runs to the building railings then back to the grass railings increase. One cat gives a quick anxious lick to a back leg, then follows with a head-over roll on the ground.
Another cat licks its back leg.

9.17p.m. Two cats run to each other, one sits down and the other one rubs its back against it as it passes.

9.18p.m. One cat raises its tail as another goes near and then rubs itself against the railings. Another couple raise tails on meeting. Another washes. Dog enters square, runs to railings and, on squeezing through, chases some of the cats. Dog chases one cat out through railings and is within inches of biting it when the cat flattens to the ground wheeling round spitting and the dog stops, then both race off around the square. The dog then urinates in the exact same spot in the gutter the previous dog had used.

9.22p.m. Mary, the cat feeder, arrives in the square, wheeling her trolley. (The cats on most evenings when undisturbed by a dog just before the feeder arrives continue with the activities of 9–9.18p.m. If the feeder is late the group will sit on the pavement staring quietly in the direction the feeder will approach from.) The cats that had run from the dog now run almost as fast back to the feeding location. Cats lift their tails in greeting to feeder and some rub around her legs and occasionally against each other similarly raising tails. One is rubbing against the railings by the grass. (The non-group tabby/white cat again appears widely skirting the group, passing along the building railings.)

9.25p.m. Dog reappears, chasing some of the cats, others stay close to Mary. But as soon as dog drifts off the cats return with tails raising in greeting. Cats feeding from the plates of food put down by Mary. Some of the cats still a little jumpy after the dog's visit, but settling down.

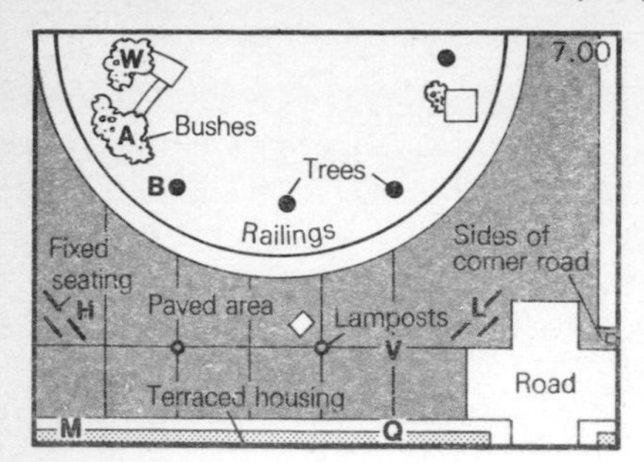
7.00
W
A
Bushes
B
Trees
Railings
Fixed seating
H
Paved area
Lamposts
Sides of corner road
L
V
Road
Terraced housing
M
Q

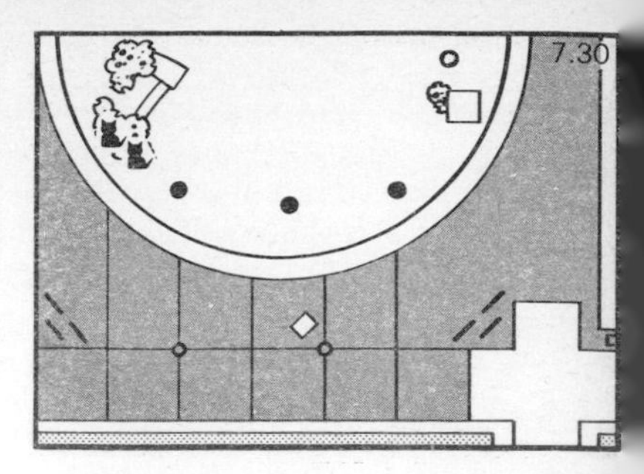
7.30

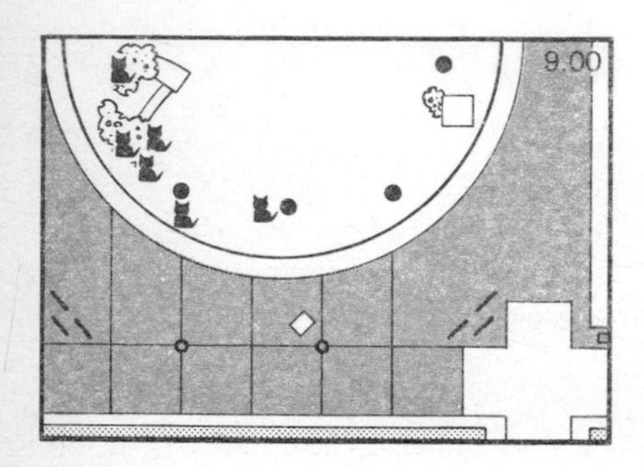
9.00

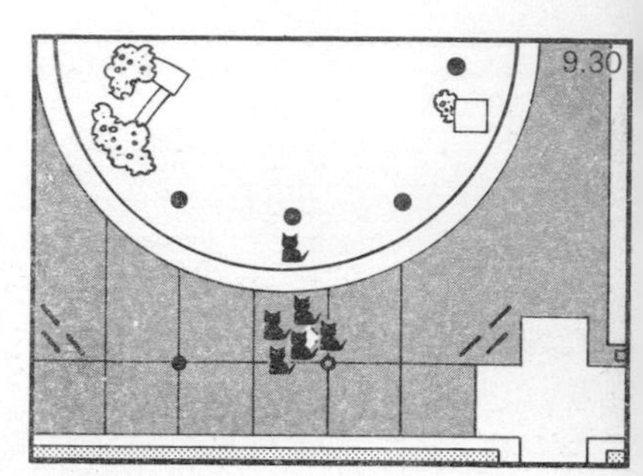
9.30

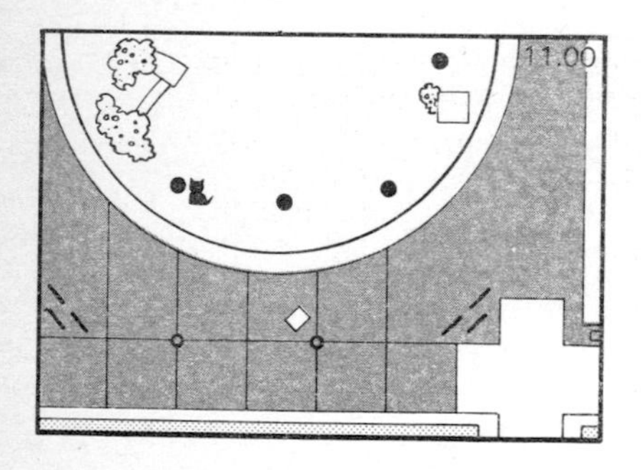
11.00

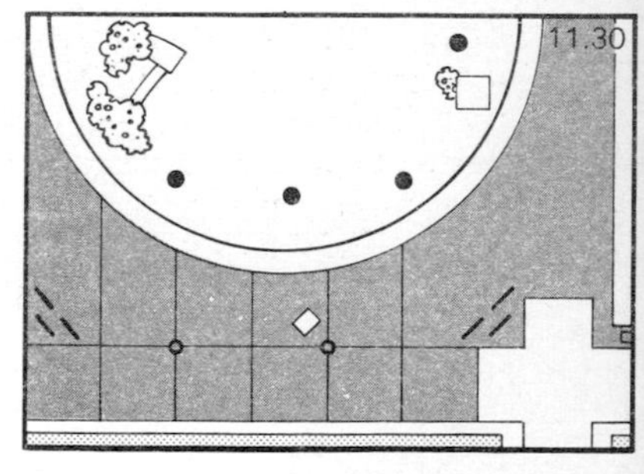
11.30

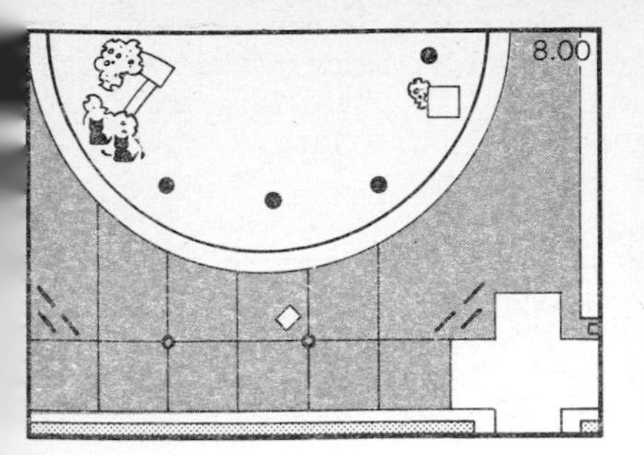

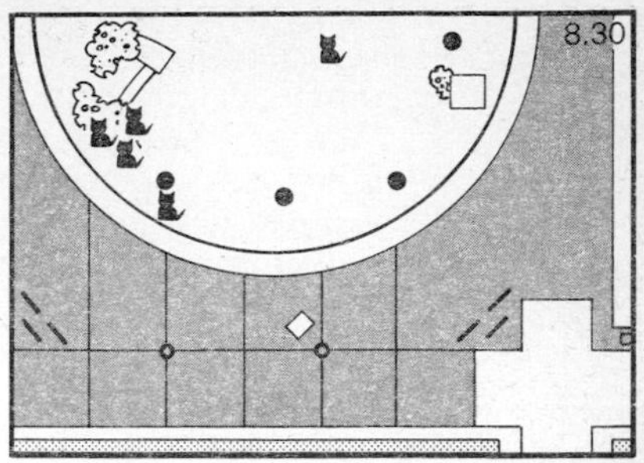

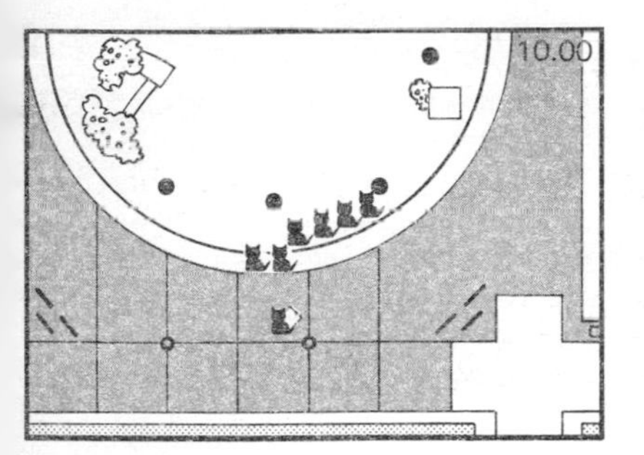

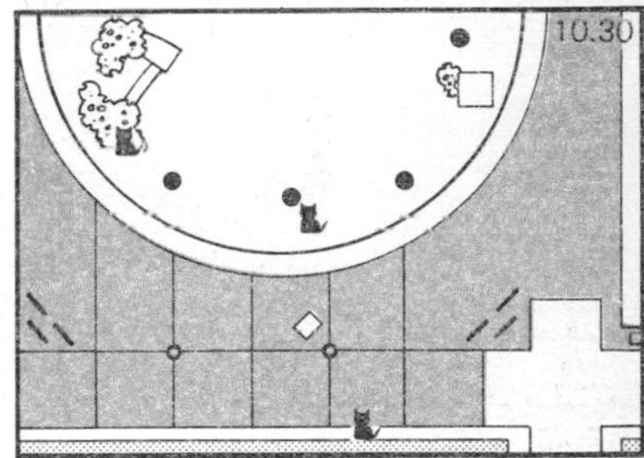

Position of the Fitzroy Square feral cats through an evening (1 June, corresponding with the evening timetables on p. 193–9 and events p. 200–1). The west half of Fitzroy Square is shown which encompasses the group's core area. The square has a railing enclosed circular grass area with trees and shrubs. The paving has a large grid of cobbled guttering. The square is bounded by tall Adam and Adam-style terraced buildings. (Key letters are those also used in timetables.)

9.30p.m. Those feeding are doing so with no aggression and heads only about six inches apart. It is only during these feeding times that the cats regularly allow the intercat distance to drop to this. Even just before and after feeding the normal minimal distance does not drop below 1–2 feet (unless specifically rubbing after tail lifting when anticipating feeding).

9.35p.m. Protective female of the group emerges from bushes (A) and approaches feeding area, hesitating at railings, but goes on to the food raising her tail as she approaches. Couple of cats stop feeding and start washing. One cat, feeding as it often does from one of Mary's containers, fishing food out with its front paw, and eating off its paw.

9.40p.m. Two cats walking along side by side, raise tails at exactly the same time. One rubs the railings, while one goes into spray position. Two more raise tails and rub railings.

9.42p.m. Six cats by Mary and two by grass railings. On average now four eating at any one time. Some occasionally washing or leisurely scratching. Suddenly one runs from the railings to Mary raising its tail – all of the others instantly start back slightly, simultaneously raising their tails vertically, but as quickly settle back down. Mary occasionally puts a hand down to stroke a cat, one or two allow this but most shy back although a number will rub around her.

9.45p.m. Tabby/white cat shows up again, walking in via the seats (L), but finds one of the cats looking at it so it shies away to the statue plinth. A little later starts to move cautiously towards the group, stopping by one of the trees. Similarly a rangy black tom not of the group slinks up to the base of another tree. Four cats still feeding at any one time.

Group cats are moving around raising tails as they go and rubbing railings by the grass. Two touch noses then rub heads.

9.50p.m. Only two left feeding – the rest of the group sitting on the plinth of the railings, equally spaced out, some washing. The strange black tom sneaks a drink of milk from a container a little away from the main feeding area. Tabby/white moves off where stared at rather fiercely by black and white (two flecks between eyes).

9.55p.m. Two feeding, most washing, one strolling across to building railings.

10.02p.m.	One cat moves off towards Virginia Woolf's house railings. One feeding, two washing, the rest just sitting quietly and still. Mary stops putting down any more food/milk.
	The strange black cat prowls about peripherially and sprays a tree.
	Another cat drifts away.
10.05p.m.	General washing, cats sitting further back towards trees.
10.10p.m.	Two more drift away across the green to be followed by another a few minutes later. One black and white still feeding. One of the group pouncing away intently following something small near the base of the statue. The strange black cat very interested in this. One returns to primary core from Virginia Woolf's railings, only to go back a few minutes later.
10.15p.m.	Two feeding, others sitting, washing (some head and shoulders washing, others feet in air, urinogenital region washing) or drifting away.
10.20p.m.	Another dog disturbs some of the cats and being a terrier, squeezes through the railings, but goes away. However as it begins to rain another cat decides to run off.
10.25p.m.	Most have now gone, couple sitting quietly and towards the plinth. The strange black cat still hovers.
10.28p.m.	Now most have gone the strange black tom goes forward, staggers slightly and sprays the other cats' food area, and then goes on to rub the grass railings with his head and then sprays one of the railing main posts. It then retraces its route sniffing cautiously and rubbing the railings hesitantly again. Only one cat of the group left and that sitting sphinx-like at the base of one of the trees. A dog comes up and eats some of the remaining food that was left by the feeder, then leaves. The black cat not intimidated by dog, but does move off soon after settling down for a wash.
10.35p.m.	Another black and white returns from bushes (A) and sniffs about, has a brief feed, then moves off to the bottom of a tree and washes. There is a fine drizzle but cats dry under trees.
10.45p.m.	Another black and white emerges and sniffs and rubs along railings plinth by bushes (A) with its head and body.
11.00p.m.	Only one cat left.
11.15p.m.	Square empty of cats.

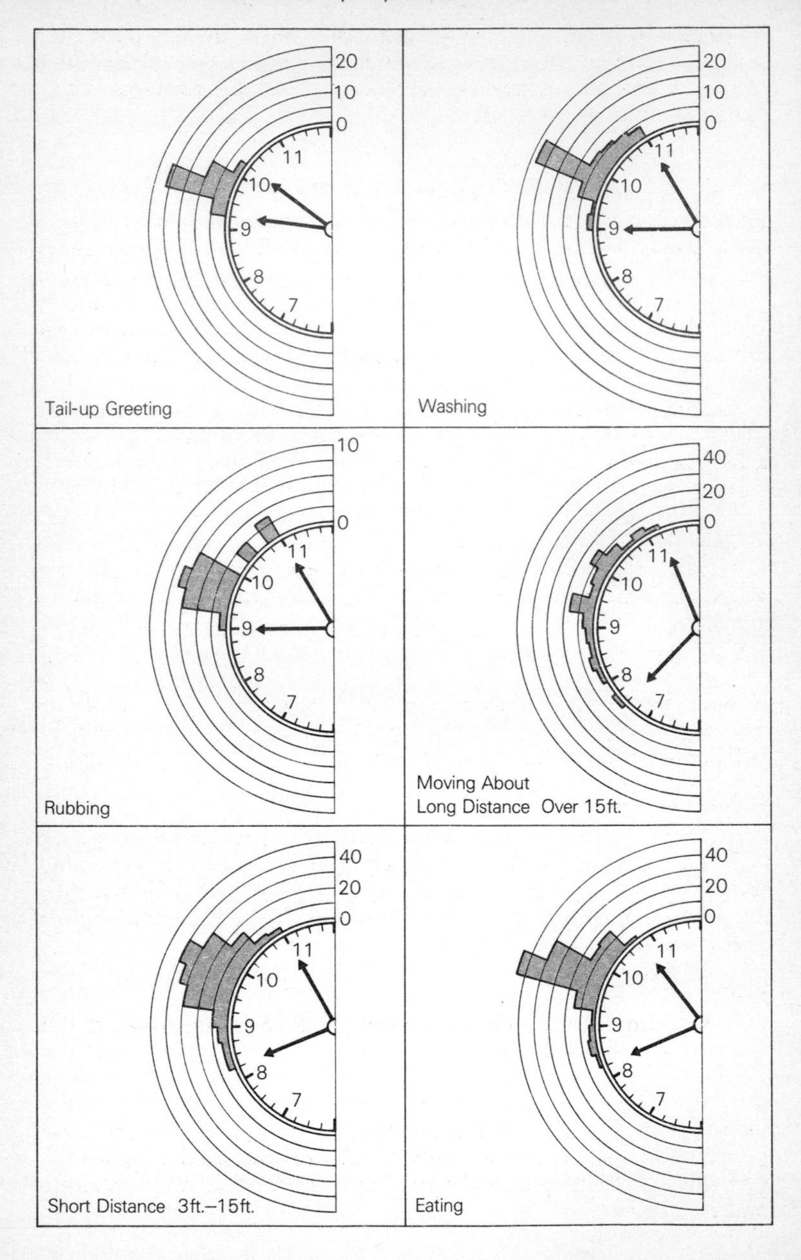
Tail-up Greeting
Washing
Rubbing
Moving About
Long Distance Over 15ft.
Short Distance 3ft.–15ft.
Eating

When recorded events for an evening are charted and then compared, the normal development of activities can be seen at a glance. (The charted events are for the same evening as the description and positions so that comparison can be made directly.)

Cat movements

The greater amount of cat movements in the general core area of the square is largely but not completely due to the greater number of cats in the square at feeding time. The cats make two distinctive moves, either for short distances, usually within the primary core area (of 3–15 feet) or long moves into or out of the primary core area, perhaps to nearby buildings (distance moves under three feet were excluded to avoid the confusion of small movements between bowls while eating – the only time such short moves regularly occur).

Coming up to feeding time more long, anxious anticipatory dashes are seen. During eating these subside, but increase after eating and as the cats leave the primary core. Short movements are at their highest during feeding periods.

Eating

The bulk of the feeding occurred over a short half-hour period, the cats only stopping when they were full. (The first small peak was due to scavenging.)

Tail-up greeting

This affectionate 'hello' greeting between group members closely follows the eating curve as cats meet for feeding. The feeder is similarly greeted by a number of cats. (Non-group members did not receive this greeting.)

Rubbing

This occurs both between cats, cats and feeder, and cats and objects in the feeding area such as railings. This mainly occurs over the same period but the later events are due to a non-group member that would like to gain admittance, leaving scent on railings after the others had largely left their feeding area.

Washing

As the cats stop eating they turn to washing before moving off (and so this curve follows upon the others).

The chart height away from the rim indicates the number of events each quarter of an hour. The hands of the 'clock' are positioned to show the span of each activity e.g. 'tail-up greeting' started at 9.15 and continued until 10.15 p.m.

Six of the Fitzroy Square feral cats showing characteristically individual black and white face markings that aided identification (still present in the square in July 1983).

Protective Queen
alternatively Big Nose Blip

Neutered female
weight (June '78): 7½lb

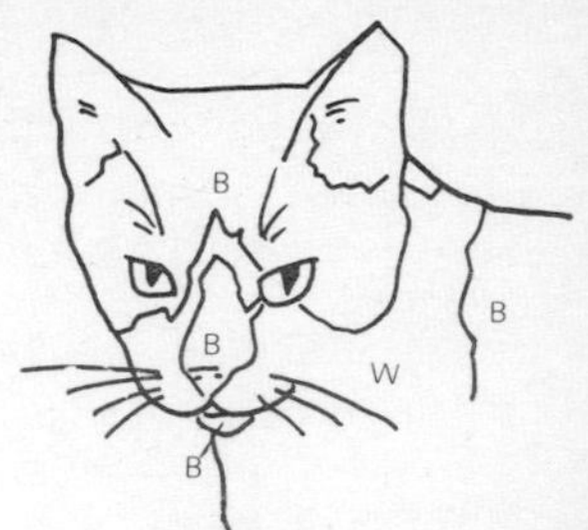

Black Nose

Neutered female
weight (June '78): 5½lb

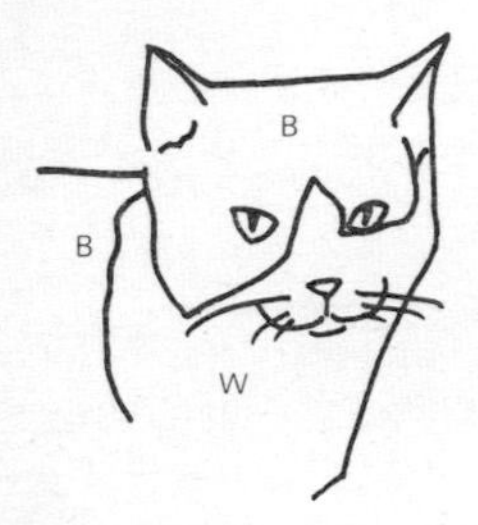

Diag

Neutered Tom
weight (June '78): 8lb

Eyebrows

Neutered female
weight (June '78): 6½lb

Double Eye Flick

Neutered female
weight (June '78): 7lb

Flat Cap

Neutered female
weight (June '78): 6lb

The black with white spotting markings also helped cat identification from body side view and head profiles, but full face was least ambiguous.

Black Nose
Neutered female
weight (June '78): 5½lb

References

HISTORICAL INTRODUCTION

DESFORGES, M. & WOODGUSH D. (1975) 'A behavioural comparison of domestic and mallard ducks; habitation and flight reactions', *Anim. Behav.*, 23, 692–7.

ELOFF, F. C. (1973) 'Ecology and behaviour of the Kalahari Lion', *Proc. 1st Conf. Ecol. Behav. and Consv. of the World's Cats*, ed. R. Eaton.

ESTES, R. D. (1972) 'The Role of the vomeronasal organ in mammalian reproduction', *Mammalia*, 36, 315–41.

ESTES, R. D. (1967) 'Predators and scavengers', *Nat. Hist. N. Y.*, 76 (2&3), 20–9, 38–47.

HALE, E. (1969) 'Domestication and the evolution of behaviour', *The Behaviour of Domestic Animals*, ed. E. Hafez, Baillière Tindall and Cassell, 2nd edn.

HARTING, J. (1878) 'Hunting the wild cat', *Zoologist Series 3*, 11, 251–2.

KIRK, J. C. & WAGSTAFFE, R. (1944) 'A contribution to the study of the Scottish wild cat', *North W Nat*, 19, 14–23.

KLEIMAN, D. & EISENBURG. J. (1973) 'Comparisons of canid and felid social systems from an evolutionary perspective', *Anim. Behav.*, 21, 637–59.

MORRISON-SCOTT, T. (1952) 'The mummified cats of ancient Egypt', *Proc. Zool. Soc. London.*, 121, 861–7.

PHILIP, J. (1974) 'The first witch trial in England', *The Witchcraft Papers*, ed. P. Haining. Robert Hale.

RADINSKY, L. (1981) 'Evolution of skull shape in carnivores', *2 Addit. Modern Carnivores. Biol. J. Linn. Soc.*, 16, 337–55.

TABOR, R. (1981) 'General biology of feral cats', *The Ecology and Control of Feral Cats*, (UFAW)*, 5–11, 47, 95.

*(UFAW) – Universities Federation for Animal Welfare.

TWICI (1876) 'Names of beasts of sport', from 'The Book of St. Albans' in *The Sports and Pastimes of the English People* by Joseph Strutt, Chatto and Windus, 75.

WEIR, HARRISON (1892) 'Our Cats', ('Our Cats and All About Them; Their Varieties, Habits and Management and for Show: the Standard of Excellence and Beauty Described and Pictured') 2nd pub. by *Fanciers Gazette Ltd* London.

WOODS, W. (1973) *A History of the Devil*, Allen.

CHAPTER 1

DARTNALL, J. A. (1975) 'Gene frequencies of feral domestic cats in Tasmania', *Carnivore Genetics Newsletter*, **2**, (9) 248–50.

ELIOT, T. S. (1940) *The Illustrated Old Possum: Old Possum's Book of Practical Cats*, Faber and Faber.

ROBINSON, R. (1959) 'Genetics of the domestic cat', *Bibliographia Genetica* **19**, 273–362.

ROBINSON, R. (1976) 'Homologous genetic variation in the Felidae', *Genetica*, **46**, 1–31.

ROBINSON. R. (1977) *Genetics for cat breeders*, Pergamon Press.

SEARLE, A. (1949) 'Gene frequencies in London's cats', *J. Genetics*, **49** pt. 3, 214–20.

SEARLE, A. (1964) 'Gene geography of cats', *J. Cat. Genetics*, **1**, 18.

TODD, N. 'Cats and Commerce', *Scientific American*.

TODD, N. (1962) 'Inheritance of the catnip response in domestic cats,' *J. Hered.*, **53**, 54–6.

TODD, N. B. & TODD, C. (1975) *Mutant allele frequences in the domestic cats of Turkey*, **2**. No. 9, 263–72.

WEIR, HARRISON, 'Our Cats', ('Our Cats and All About Them: Their Varieties, Habits and Management and for Show, the Standard of Excellence and Beauty Described and Pictured'). 2nd Pub. by *Fanciers Gazette Ltd.*, London. 1892.

CHAPTER 2

DAW, N. W. & PEARLMAN, A. (1970) 'Cat colour vision: evidence for more than one cone process', *J. Physiol*, **21**, 1125–37.

GUNTER, R. (1951) 'The absolute threshold for vision in the cat', *Phys.*, **114**, 8–15.

LEYHAUSEN, P. (1979) *Cat Behaviour*, Garland STPM Press.

NEFF, W. D. & HIND, J. E. (1955) 'Auditory thresholds of the cat', *J. Acoust. Sov. Amer.*, 27, 480–3.

ULMER, M. J., HAUPT, R. E. & HICKS, A. (1971) *Anatomy of the cat – an atlas and dissection guide*, Harper and Row.

CHAPTER 3

LEYHAUSEN, P. (1979) *Cat Behaviour*, Garland STPM Press.

MOELK, M. (1944) 'Vocalizing in the house-cat; a phonetic and functional study', *Am. J. Psych.*, 57, 184–205.

CHAPTER 4

BREARLEY, E. & KENSHALO, D. (1970) 'Behavioural measurements of the sensitivity of the cat's upper lip to warm and cool stimuli', *J. Compar. and Physiol. Psy.*, 70, 1.

INWARDS, R. (1950) *Weather Lore*, SR Publishers Ltd, 4th edn.

CHAPTER 5

BARON, A., STEWART, C. N. & WARREN, J. M. (1957) 'Patterns of social interaction in cats' *Behaviour*, 11, 1, 56–66.

BURT, W. (1943) 'Territoriality and home range concepts as applied to mammals', *J. Mammal.*, 24, 346–52.

COLE, D. D. & SHAFER, J. N. (1966) 'A study of social dominance in cats', *Behaviour*, 27, 1, 39–53.

DARDS, J. (1978) Home ranges of feral cats in Portsmouth Dockyard. Proc. 1st Int. Conf. Domestic Cat Population Genetics & Ecology. *Carnivore Genetics Newsl.*, 3, 242–55.

DARDS, J. (1981) 'Habitat utilisation by feral cats in Portsmouth Dockyard', *The Ecology and Control of Feral Cats* (UFAW) 30–46.

HEDIGER, H. (1950) *Wild animals in captivity*, Butterworth. Sc. Pub.

LAUNDRE, J. (1977) 'The daytime behaviour of domestic cats as a free-roaming population', *Anim. Behav.*, 25, 4, 990–8.

LEYHAUSEN, P. (1965) 'The communal organisation of solitary mammals', *Symp. Zool. Soc. London*, 249–63.

LEYHAUSEN, P. (1979) *Cat Behaviour*, Garland STPM Press.

MACDONALD, D. (1981) 'The behaviour and ecology of farm cats', *The Ecology and Control of Feral Cats*, (UFAW), 23–9.

MACDONALD, D. & APPS (1978) 'The social behaviour of a group of semi-dependent farm cats, Felis Catus; a progress report', *Proc. 1st Int. Conf. Domestic Cat Population Genetics and Ecology. Carnivore Genetics Newsl.*, 3, 256–68.

MASSERMAN, J. & SIEVER, P. (1944) 'Dominance, neurosis and aggression; an experimental study', *Psychosomatic Med.*, 6, 7–16.
MCNAB, B. (1963) 'Bioenegetics and the determination of home range size', *Amer. Nat.*, 97, 133–40.
TABOR, R. (1981) 'General biology of feral cats', *The Ecology and Control of Feral Cats* (UFAW), 5–11, 47, 95.
WINSLOW, C. (1938) 'Observations of dominance – subordination in cats', *J. Genetic. Psy.*, 52, 425–8.

CHAPTER 6

HOWES, C. 'Personal communication'.
HOWES, C. (1976) 'The ecology of the Yorkshire mog', *Yorkshire Naturalists' Union Newsl.*, 36.
HOWES, C. (1980) 'The great British Cat Survey', *Mammal Soc. Youth News*, 12.
HOWES, C. (1979) 'What the cat brought in', *Mammal Soc. Youth News*, 7.
MATHESON, C. (1944) 'The domestic cat as a factor in urban ecology', *J. Amer. Ecol.*, 13, 130–3.

CHAPTER 7

COMAN, B. & BRUNNER, H. (1972) 'Food habits of the feral house cat in Victoria', *J. Wildl. Mangmt.*, 36(3), 848–53.
EBERHARD, T. (1954) 'Food habits of Pennsylvania house cats', *J. Wildl. Mangmt.* 18, 284–6.
ENGLUND, J. (1965) 'Studies on food ecology of the red fox (*Vulpes vulpes*) in Sweden', *Viltrovy (Swedish Wildlife)*, 3, (4).
ENVIRONMENT, Dept. of. (1975) *Refuse disposal*, HMSO.
ENVIRONMENT, Dept. of. (1975) *Refuse storage and collection*, HMSO.
ERRINGTON, P. (1936) 'Notes on food habits of South Wisconsin house cats', *J. Mammalogy.*, 17, (1), 64–5.
HAMILTON, W. & HUNTER, R. (1939) 'Fall and winter food habits of Vermont bobcats', *J. Wildl. Mangmt*, 3, 99–103.
HARRIS, S. (1981) 'The food of suburban foxes (*Vulpes vulpes*) with special reference to London', *Mammal Rev.*, 11, (4), 151–68.
HUBBS, E. (1951) 'Food habits of feral house cats in the Sacramento Valley', *Calif. Fish & Game*, 37, 177–89.
JOSHUA, J. (1965) 'The clinical aspects of some diseases of cats', Heinemann.
LLEWELLYN, L. & UHLER, F. (1952) 'The foods of fur animals of the Patuxent research refuge, Maryland', *Amer. Midl. Nat.*, 48, 193–203.

MACDONALD, D. (1977) 'On food preferences in the red fox', *Mammal Rev.*, 7, No. 1, 7–23.
MCMURRY, F. & SPERRY, C. (1941) 'Food of feral house cats in Oklahoma', *J. Mammology*, **22**, 185.
PULLAR, P. *Sunday Times Magazine*.
REMFRY, J. 'Personal communication'.
SCOTT, P., CARVALHO DA SILVO, A., & LLOYD-JACOB, M. (1957) 'The Cat', ch. 45 UFAW handbook on the care and management of laboratory animals, ed. A. N. Worden and W. Lane-Petter, 2nd edn.

CHAPTER 8

BARRETT, P. & BATESON, P. (1978) 'The development of play in cats', *Behaviour*, **66**, 106–20.
BIBEN, M. (1979) 'Predation and predatory play behaviour of domestic cats', *Anim. Behav.*, **27**, 81–94.
BRADT, G. W. (1949) 'Farm cat as predator', *Michigan Conserv.*, **18**, pt. (4), 23–5.
CARO, T. M. (1981) 'Sex differences in the termination of social play in cats', *Animal Behaviour*, **29**, 271–9.
COMAN, B. & BRUNNER, H. (1972) 'Food habits of the feral house cat in Victoria', *Wildl. Mangmt*, **36** (3), 848–53.
CROWCROFT, P. (1954) 'The daily cycle of activity in British shrews', *Proc. Zool. Soc. London.*, **123**, 715–29.
EBERHARD, T. (1954) 'Food habits of Pennsylvania house cats', *J. Wildl. Mangmt*, **18**, 284–6.
ELTON, C. (1953) Cats in farm rat control. *Brit. J. Animal Behav.*, **1**, 151–5.
ERRINGTON, P. (1936) 'Notes on food habits of South Wisconsin house cats', *J. Mammalogy*, **17**, (1), 64–5.
FITTER, R. S. R. (1945) *London's Natural History*, Collins.
FITTER, R. (1968) *The vanishing world of animals*, Midland Bank in association with Kaye and Ward Ltd.
HOWES, C. 'Personal communication'.
HOWES, C. (1976) 'The ecology of the Yorkshire Mog', *Yorkshire Naturalists' Union Newsl.* **36**.
HOWES, C. (1980) 'The Great British Cat Survey', *Mammal Soc. Youth News*, **12**.
HOWES, C. (1979) 'What the cat brought in', *Mammal Soc. Youth News*, **7**.
HUBBS, E. (1951) 'Food habits of feral house cats in the Sacramento Valley', *Calif. Fish and Game*, **37**, 177–89.
KUO, Z. Y. (1931) 'The genesis of the cat's responses towards the rat', *J. Comp. Psy.*

LEYHAUSEN, P. (1973) 'On the function of the relative hierarchy of moods – as exemplified by the phylogenic and ontogenic development of prey-catching in carnivores', *Motivation of Human and Animal Behaviour: an Ethological View*, ed. K. Lorenz and P. Leyhausen, Van Nostrand Reinhold.

LLEWELLYN, L. & UHLER, F. (1952) 'The foods of fur animals of the Patuxent research refuge, Maryland', *Amer. Midl. Nat.*, **48**, 193–203.

LOCKLEY, R. (1964) *The private life of the rabbit*, Andre Deutsch.

MACDONALD, D. & APPS, (1978) 'The social behaviour of a group of semi-dependent farm cats, Felis catus; a progress report', *Proc. 1st Int. Conf. Domestic Cat Population Genetics and Ecology, Carnivore Genetics Newsl.*, 3, 256–68.

MCMURRY, F. (1945) 'Three shrews eaten by a feral house cat', *J. Mammology*, **26**, 94.

MCMURRY, F. & SPERRY, C. (1941) 'Food of feral house cats in Oklahoma', *J. Mammology*, **22**, 185.

MEAD, C. (1982) 'Ringed birds killed by cats', Mammal. Rev. **12**(4), 183–6.

NADER, I. & MARTIN, R. (1962) 'The shrew as prey of the domestic cat', *J. Mammalogy*, **43**, 417.

PARMALEE, P. (1953)'Food habits of the feral house cat in east-central Texas', *J. Wildl. Mangmt*, **17** (3), 375–6.

RANSOME, R. (1980) *The greater horseshoe bat*, Blandford Press, Mamm. Soc.

TONER, G. (1956) 'House cat predation on small mammals', *J. Mammalogy*, **37** 1, 119.

TSAI, L. S. (1963) 'Peace and co-operation among "natural enemies" educating a rat killing cat to co-operate with a hooded rat', *Act. Psychol. Taiwan*, **5**, 1–5.

WALTON, C. (1947) *Farmers warfare*, Grosby Lockwood & Son Ltd.

WILSON, G. & WESTON, E. (1946) *The cats of wildcat Hill*, Duell, Sloan & Pearce.

CHAPTER 9

BARRETT, P. & BATESON, P. 'The development of play in cats', *Behaviour*, **66**, 106–20.

BLOOM, F. (1960) *The urine of the dog and cat – analysis and interpretation*, Gammon Pub. Inc. NY.

CORBETT, L. K. (1978) 'A comparison of the social organisation and feeding ecology of domestic cats in two contrasting environments in Scotland', *Carnivore Genetics Newsl.*, **3**, 269.

CORBETT, L. K. (1979) 'Relationships between the feeding ecology and social organisation of cats', *NERC/ITE Annual Report.*

ESTE, R. D. (1972) 'The role of the vomeronasal organ in mammalian reproduction, *Mammalia*, **36**, 315–41.

EWER, R. F. (1959) 'Suckling behaviour in kittens', *Behaviour*, **15**, 146–62.

JOHNSON, R. (1973) 'Scent marking in mammals', *Anim. Behav.*, **21**, 521–35.

KOLB, B. & NONEMANN, A. (1975) 'The development of social responsiveness in kittens', *An. Behav.*, **23**, 368–74.

LEYHAUSEN, P. (1965) 'The communal organization of solitary mammals', *Symp. Zool. Soc. of London*, 249–63.

LEYHAUSEN, P. (1979) *Cat Behaviour*, Garland STPM Press.

MYKYTOWYCZ, R. (1970) 'The role of skin glands in mammalian communication', *Communication by Chemical Signals, Advances in Chemoreception*, 1 Ch. 11, ed. Johnston J, Moulton, D, & Turk, A. Appleton, Century & Crofts.

PALEN, C. & GODDARD, G. (1966) 'Catnip and oestrous behaviour in the cat', *Anim. Behav.*, **14**, 372–77.

SCHAFFER, J. (1940) *Die Hautdrusenorgane der Saugetiere*, Urban & Schwarzenberg.

TODD, N. B. (1962) 'Inheritance of the catnip response in domestic cats', *J. Hered.*, **53**, 54–6.

CHAPTER 10

JACKSON, M. 'Personal communication'.

LATIMER, H. (1937) 'The weights of linear measurement of the digestive system of the adult cat', *Anat. Record*, **68**, 469–80.

LATIMER, H. (1936) 'Weights and linear measurements of the adult cat', *Am. J. Anat.*, 58, No. 2, 329–47.

PHIPSON, J. (1976) *The Cats*, Macmillan.

WEIR, HARRISON. (1892) 'Our Cats', ('Our Cats and All About Them: Their Varieties, Habits and Management and for Show: the Standard of Excellence and Beauty Described and Pictured') 2nd pub. by *Fanciers Gazette Ltd* London.

CHAPTER 11

ANDERSON, G. AND CONDY, P. (1974) 'A note on the feral house cat and house mouse on Marion Island', *S. African J. Antarctic Res.*, **4** 58–61.

ASHMOLE, N. (1963) 'The biology of the Wideawake or Sooty Tern on Ascension Island', *Ibis*, *1036*, 297–365.

BUDRIS, J. 'Personal communication'.

DIXON, D. 'Personal communication'.

KONECNY, M. 'Personal communication'.

STONEHOUSE, B. (1962) 'Ascension Island and the British Ornithologists' Union Centenary Expedition 1957–9, *Ibis*, *1036*, 107–23.

TAYLOR, R. H. (1979) 'Predation on sooty terns at Raoul Island by rats and cats', *Notornis*, **26**, 199–202.

THORNTON, I. (1971) *Darwin's Islands, a Natural History of the Galapagos*, Amer. Mus. Nat. Hist./Nat. Hist. Press.

WILLIAMS, A. J. (1980) 'Aspects of the breeding biology of the subantarctic Skua at Marion Island', *Ostrich*, **51**, 160–7.

CHAPTER 12

ALCOCK, J. 'Tuesday Call', BBC Radio 4 Broadcast, August 1977.

ARUNDEL, J. (1970) 'Control of helminth parasites of dog and cat', *Aust. Vet. J.*, **41**, 164–8.

AVERY, R. A. (1974) 'Parasites of dogs and cats', *Identification of animal parasites*, Hutton Group Keys.

BEARUP, A. (1960) 'Parasitic infection in cats in Sydney, with special reference to the occurrence of *ollulanus tricuspis*', *Aust. Vet. J.*, **36**, 352–4.

BECK, A. (1973) *The Ecology of stray dogs – a study of fee-ranging urban animals*, York Press, Baltimore.

BISSERU, B. (1967) *Diseases of man and his pets*, Heinemann.

BISSERU, B. (1967) 'Diseases of man acquired from his pets – *Toxocara* infections', *Carnivores*, 90–104, Heinemann.

BORG, O. & WOODRUFF, A. (1973) 'Prevalence of infective ova of *Toxocara* species public places', *Bri. Med. J.*, **4**, 470–2.

COMAN, B. (1972) 'A survey of the gastro-intestinal parasites of the feral cat in Victoria', *Aust. Vet. J.*, **48**, 133–6.

COMAN, B. & BRUNNER, H. (1972) 'Food habits of the feral house cat in Victoria', *J. Wildl. Managmt*, **36**. (3), 848–53.

COMFORT, A. (1956) 'Maximum ages reached by domestic cats', *J. Mamm.* 37, 118.

DUBEY, J. (1966) '*Toxocara cati* and other intestinal parasites of cats', *Vet. Rec.*, **79**. (18), 506–8.

GALLIARD, H. (1974) '*Larva migrans* – from parasitic zoonoses', *Clinical Experimental Studies*, ed. E. Soulsby.

GIBSON, T. E. (1960) '*Toxocara canis* as a hazard to public health', *Vet. Rec.*, 72, 772–4.

GRAHAM-JONES, O. (1966) 'Some diseases of animals communicable to man in Britain', *Proc. of Symposium: Bri. Vet. Ass. and Bri. Sm. An. Vet. Ass.*, Pergamon Press.

GREGORY, G. & MUNDAY, B. (1976) 'Internal parasites of feral cats from the Tasmanian Midlands and King Island', *Aust. Vet. J.*, **52**, 317–20.

HAMILTON, J., HAMILTON, R. & MESTLER, G. (1969) 'Duration of life and causes of death in domestic cats influence of sex, gonadectomy and inbreeding', *J. Gerontology*, **24**, 427–37.

HOWES, C. (1980) 'The Great British Cat Survey', *Mammal Soc. Youth News*, **12**.

JOHNSTON, D. H. & BEAUREGARD, M. (1969) 'Rabies epidemiology in Ontario', *Wildf. Dis. Ass.*, **5**.

KEEP, J. (1970) 'Gunshot injuries to urban dogs and cats', *Aust. Vet. Record*, **46**, 330–4.

KOLATA, R., KRAUT, N. & JOHNSTON, D. (1974) 'Patterns of trauma in urban dogs and cats: a study of 1,000 cases', *J. A. M. Vet. Med. Assoc.*, **164**, 499–502, **11**, 1–35.

LAMINA, J. (1974) 'Immunodiagnosis of visceral larval migrans in man for parasitic zoonoses', *Clinical and Experimental Studies*, ed. E. Soulsby.

NIAK, A. (1972) 'The prevalence of *toxocara cati* and other parasites in Liverpool cats', *Vet. Rec.*, **91**, 534–6.

NICHOLSON, G. (1981) 'Government policy on rabies control', *The Ecology and Control of Feral Cats*, (UFAW), 50–9.

OLDHAM, J. (1965) 'Observations on the incidence of *Toxocara* and *Toxascaris* in dogs and cats from the London area', *J. Helmintology*, **39**. (2/3), 251–6.

RYAN, G. (1976) 'Gastro-intestinal parasites of feral cats in New South Wales', *Aust. Vet. J.*, **52**, 224–7.

SLOAN, J. (1961) *Advances in small animal practice*, **2**, Pergamon Press.

SPRENT, J. F. (1956) 'The life history and development of *Toxocara cati* (Shrank 1788) in the Domestic Cat', *Parasitotology*, **46**, 1, 2, 54–78.

THOMSETT, J. (1963) 'Diseases transmitted to man by dogs and cats', *The Practitioner*, **191**, 630–40.

WOODRUFF, A., THALKER, C. & SHAH, N. (1964) 'Infection with animal helminths', *Br. Med. J.*, **1**, 1001–5.

CHAPTER 13

ARONSON, L. & COOPER, M. (1967) 'Penile Spines of the domestic cat: their endocrine behaviour relations', *Anat. Rec.*, **157**, 71–8.

GREULICH, W. W. (1934) 'Artificially induced ovulation – the cat (*Felis domestica*)', *Anat. Rec.*, **58**, 217–24.

HAMILTON, J., HAMILTON, R. & MESTLER, G. (1969) 'Duration of life and causes of death in domestic cats: influence of sex, gonadectomy and inbreeding', *J. Gerontology*, **24**, 427–37.

HARRIS, T. & WOLCHUK, N. (1963) 'The suppression of oestrus in the dog and cat', *Am. J. Vet. Res.*, **24**, 1003.

HEDIGER, H. (1950) *Wild animals in captivity*, Butterworth Sc. Pub.

JACKSON, M. 'Personal communication'.

MANOLSON, F. (1971) *My cat's in love*, Pelham Books.

MICHAEL, R. (1961) 'Observations upon the sexual behaviour of the domestic cat. (*Felis catis*) under laboratory conditions', *Behaviour*, **18**.

RETTER, E. (1887) 'Effets de la castration sur l'évolution des tissus peniens chez le chat', *C. R. Soc. Biol.*, **39**, 206–8.

RETTER, E. & LE LIEVRE, P. (1912) 'Effets de la castration sur le chat', *C. R. Soc. Biol.*, **2**, 184–6.

ROSENBLATT, J. & ARONSON, J. (1958) 'The influence of experience on the behavioural effects of androgen on prepuberty castrated male cats', *Animal Behaviour*, **6**, 171–82.

SCOTT, P. & LLOYD-JACOB, M. (1955) 'Some interesting features in the reproduction cycle of the cat, *Studies in Fertility*, **7**, 123–9.

SCOTT, P., CARVALHO DA SILVA, A., & LLOYD-JACOB, M. (1957) 'The Cat', *UFAW handbook on the care and management of laboratory animals*, Ch. 45, ed. A. N. Worden & W. Lane-Petter. 2nd edn.

CHAPTER 14

DARDS, J. (1981) 'Habitat utilisation by feral cats in Portsmouth Dockyard', *The Ecology and control of feral cats*, (UFAW), 30–46.

GRODZINSKI, W., PUCEK, Z. & RYSZKOWSKI, L. (1966) *Acta. theriol.*, **11**, 297.

KIRSTENSEN, T. (1981) 'Feral cat control in Denmark', *The Ecology and Control of Feral Cats*, (UFAW), 68–72.

JACKSON, M. 'Personal communication'.

JACKSON, M. (1981) 'Professional trapper's view', In *The Ecology and Control of Feral Cats*, (UFAW), 92–94.

PEDIGREE PET FOODS, *Pets and the British*, 3rd edn., 1979.

REMFRY, J. 'Strategies for control', *The Ecology and Control of Feral Cats*, (UFAW), 73–9.

TABOR, R. (1981) 'General biology of feral cats', *The Ecology and Control of Feral Cats*, (UFAW), 5–11, 47, 95.

YOUNG, S. (1981) 'Colonies in Hospitals', *The Ecology and Control of Feral Cats*, (UFAW), 83–5.

Acknowledgements

Over the six years that this book was researched and written a number of studies have produced data sufficient to form a better understanding from new viewpoints of an animal that we have lived alongside since its recognized birth as a species. The work of Dr David MacDonald and Peter Apps and Dr Jane Dards has in particular contributed to this re-evaluation. Among earlier workers the studies of Paul Leyhausen must be considered seminal. I must thank Colin Howes of the Doncaster Museum and Art Gallery and E. Jane Ratcliffe for making their north of England studies available to me. Similarly I am most grateful to both Mike Konecny of Florida University Zoology Department and John Budris of Massachusetts for sending me data and photographs of Galapagos Island cats. (Pictures of Galapagos Islands' feral kitten, and Galapagos Hawk with a cat as prey are both by John Budris.) Following protracted correspondence across the Atlantic with Mike, it was both a great pleasure and a useful exchange of information when by pure chance I met him on the Galapagos in 1981. I was returning to the islands leading a party from Britain, and he in helping a stranger land was the first person I met on Santa Cruz Island. Neither had reason to recognize the other, and it was only when conversation revealed we both studied feral cats that realization dawned who the other was, and that we were already well acquainted by letter!

In the course of my own field and survey researches that have contributed to the book, I have been kindly helped by more people than I can possibly mention, and anyone I cannot squeeze in I hope will none the less accept my thanks. I have received more practical help from Mike Jackson than anyone else. He and his staff at Jackson and Booth Services gave freely of their time and expertise for which I am most appreciative. I must thank the many feral cat-feeding ladies that have co-operated with me, and in particular Miss Mary Wyatt who was for many years the Fitzroy Square feeder (until prevented from continuing by a road accident). Mary was fully involved with the neutering programme at the square, and cared deeply for her cats. Edie who also fed adjacent groups has sadly recently died. Mr A. Budden and Mrs Doris Westwood of the Fitzroy Square Frontagers and Garden Committee from the beginning kindly made the square available for the study. Much of the attractive nature of the square has derived from Mr Budden and his committee, and in keeping their attitude to the square's independent living cats has been consistently enlightened.

The Cat's Protection League helped with care and assistance for the cats. I would like to thank Mrs De Clifford, Major Garforth and Group Captain Boothby and staff. I was particularly grateful for the facilities and help extended to me by Mr and Mrs Richardson also of the Cats' Protection League at Battlesbridge in Essex. I am also grateful to members of the Cat Action Trust (which I had the privilege to help into existence), particularly Miss Ruth Plant, its driving spirit, and Miss Celia Hammond, also members of the former National Cat Rescue Co-ordinating Committee. At my Barking study site I would like to thank all those who co-operated, and in particular Mr and Mrs Sibley and Mr and Mrs Edwards. At the North East London Polytechnic I would like to thank parasitologists Dr Stan Ball, Dr Keith Snow and Sue Nichol, and for keeping the mobile unit mobile Dr Peter Spence and his assistant Tom Bateman; also Norman Weedon. Special thanks to Mrs Joan Hodge of the Electron Microscope suite, (plus Dr John Wright and Maurice Watson for their treasured tolerance!)

My thanks to George Inger, ex-producer of BBC 1's 'Animal Magic' for letting me stage the cat ear-washing survey. I would also like to thank Roy Robinson; the Danish vet, Tom Kristensen; North East London Polytechnic and Linnean Society Library staff for their considerable help; Jules Azzopardi; the many kind people writing to me with site, feral taming and other information on cats of their acquaintance; Carol Haspel of New York City University Biology Department; Reg Smith of Weyhill Wildlife Park; Dr Graham Gregory, the Veterinary Officer (Vertebrate Pests) Department of Agriculture, Tasmania; Dr Jenny Remfry and Roger Ewbank of the Universities Federation for Animal Welfare; Michael Bentine and his family, including the aged family cat (colour picture 1.2); Rex Shepard of the British Museum and his committee; the *Evening Standard*; staff and students of the North Havering Adult Education College; David Dixon; and Bryn Jervis-Read.

I must thank Dr Keith Snow and veterinarian Frank Brock for reading parts of the manuscript, but any errors, particularly in the parts they didn't read, fall back on me (or the publisher!). I would like to thank Hutchinson and Arrow for their helping to promote some of the original research included in the book and for their tolerant handling of the project, in particular Terence and Caroline Blacker, Jane Judd and Roger Walker. I really ought to thank Richard Mabey for asking me to write the book in the first place, but as the time spent on writing a book is not all unmitigated joy he will understand if they are slightly qualified thanks!

Finally, beyond anyone else, I would like to thank my wife, Rachel, without whom this book would indeed most certainly not have been written. Her continuous and considerable support has made it all possible.

I have been delighted in the acquaintance of many cats during the course of my researches and the writing of this book – and it is not a flippance to include them in the acknowledgements, particularly our own house cat.

Roger Tabor, 1983

Index